Natural Gas Liquids

NATURAL GAS LIQUIDS

A NONTECHNICAL GUIDE

WILLIAM L. LEFFLER

PennWell®

Disclaimer

The recommendations, advice, descriptions, and the methods in this book are presented solely for educational purposes. The author and publisher assume no liability whatsoever for any loss or damage that results from the use of any of the material in this book. Use of the material in this book is solely at the risk of the user.

PennWell Corporation
1421 South Sheridan Road
Tulsa, Oklahoma 74112–6600 USA

800.752.9764
+1.918.831.9421
sales@pennwell.com
www.pennwellbooks.com
www.pennwell.com

Marketing Manager: Sarah DeVos
National Account Executive: Barbara McGee

Director: Mary McGee
Managing Editor: Stephen Hill
Production Manager: Sheila Brock
Production Editor: Tony Quinn
Book Designer: Susan E. Ormston
Cover Designer: Karla Womack

Library of Congress Cataloging-in-Publication Data

Leffler, William L.
Natural gas liquids : a nontechnical guide / William L. Leffler.
pages cm
Includes index.
ISBN 978-1-59370-324-0
1. Liquefied natural gas. 2. Gas as fuel. 3. Liquefied natural gas industry. I. Title.
TP761.L5L38 2014
665.7'73--dc23

2014013639

Printed in the United States of America

2 3 4 5 18 17 16 15

Contents

Foreword

Natural gas liquids have always occupied odd corners of the oil and gas industry. They didn't fit easily into the upstream or the downstream. And they still don't. That's why a good treatment of this obscure set of commodities demands someone who has worked in all parts of the value chain—E&P, refining, marketing, and petrochemicals. That person is Bill Leffler, who spent 36 years at Shell in all these sectors, even as Gas Liquids Business Manager for years.

I worked with Bill for many years. I can think of no one better to translate the complexities of natural gas liquids into a more easily understandable subject.

I recommend this book from front to back if you want to understand how ethane, propane, butane, and natural gasoline touch all parts of our business in a profound and sometimes subtle way.

Frank H. Richardson
President and CEO, Shell Oil Company, Ret.

Preface

The first textbook ever, written by Peter Lombard, was published in 1158. In all the years that have followed, incredibly enough, no one has put a pen to paper to create another textbook explaining natural gas liquids (NGLs) in a simple way. This work corrects that incomprehensible void.

Just a few decades ago, most engineers and businesspeople who managed and operated NGLs inhabited offices that were situated in the back corridors of oil companies' offices or in obscure buildings out in the oil patch. Their more senior executives thought of their commodities as necessary but ancillary operations to the much larger oil and gas businesses.

In more recent decades, three waves of change rolled in to alter most of that:

- The emergent and re-emergent growth of the petrochemicals industry, which provided excellent feedstock opportunities for NGLs.
- The appearance of dynamic and entrepreneurial midstream companies whose primary business is gas plants, pipelines, storage, and any other logistical matter. Exploration and production companies were happy to outsource their midstream worries to them.
- The burgeoning supply of NGLs, especially from shale gas exploitation.

So today, numerous enterprises populate the value chain from the subsurface reservoir to the consumer. To deal with that, I am fortunate to have had the opportunity to tap into people with many different areas of expertise. In particular, I want to thank Dr. Shu Luo, Dr. Donald L, Burdick, Martin S. Raymond, Joel Sopchak, Josh Kasprzak, Gary Hunter, Michael Tusiani, and Robert R. Petrie. They all helped refine the ideas and words I have used to explain the NGL business in terms that are as nontechnical as possible. Still, they have no responsibility and get no reward (other than their name in print) for the integrity of anything in this book. I get both.

Not every chapter in this book will appeal to someone looking for a better understanding of NGLs. Certainly, the introduction and discussion of the value chain in chapter 1 and the history of the NGL business in chapter 2 should appeal to all. The science and engineering of the upstream in chapter 3 might be a dauntingly deeper background (pun intended) than some want. As technical as the chemistry and physics of NGLs is in chapter 4, it discusses a promisingly important foundation of the business and should not be missed.

The rest of the chapters, on gas plants (chap. 5), refineries (chap. 6), logistics (chap. 7), petrochemical markets (chap. 8), and fuel markets (chap. 9), speak to understanding the commercial aspects of NGLs. Chapter 10 covers safety, operations, and maintenance. Profiles of some intriguing NGL companies in chapter 11 address the elements of success in the NGLs business. For anyone even mildly into technology, the appendices give some particulars that elaborate on the exasperatingly complex problems that NGLs can present to the owners of natural gas pipelines and gas plants and to their customers.

William L. Leffler
January 2014

CHAPTER 1

Introduction

The beginning is the most important part of the work.

Plato (427–347 BCE), *The Republic*

In 1910, a frustrated motorist complained to Dr. Walter Snelling of the U.S. Bureau of Mines, "Half the gas in my tank disappears by the time I get home." Snelling took a sample and quickly found the disappearing liquid to be vaporizing gases. He built his own still from an old water heater, some copper tubing, and wire-wrapped glass bottles. That allowed him to collect the gases as liquid and to patent the first process for capturing natural gas liquids. Two years later, Snelling's American Gasol Company installed a propane tank at the home of John Gahring, his first commercial customer, which marked the start of the liquefied petroleum gas industry.

Snelling was an exceptional scientist but not a businessman. His ventures collapsed, and in 1913 he sold his patent to the more entrepreneurial Frank Phillips for $50,000, a handsome sum at the time, and returned to the Bureau of Mines.

That oxymoron, natural gas liquids, can be a mystery. Are they gases? Are they liquids? Are they natural?

The answer to all three questions: yes . . . usually.

In the middle of the 20th century natural gas liquids (NGLs) operations were an obscure business run in the back corridors of most exploration and production companies, down the hall from the big traders. Later, the growth of the petrochemical industry and the development of transportation technologies lent more stature to the liquids crowd. More recently, the burgeoning supply around the world has put NGLs high on business agendas. The logistics of natural gas and NGLs has become focused enough that it has taken on its own *persona* as part of the midstream, which includes the pipelines, trading, storage, and other operations that fit in between the upstream and downstream.

The Value Chain

The NGLs—ethane, propane, butane, and natural gasoline—come from natural gas wells or from associated gas from oil wells. In either case, the gas streams are separated from oil at the well site. (At a gas well, the oily fluid is called condensate; at an oil well, it is called crude oil.) The gas then typically goes to a gas plant. Gas plants clean up contaminants in the gas and NGLs; they separate NGLs from the natural gas; and fractionators separate the NGLs from each other. The various NGLs, each meeting its own specification, move into several shared markets: refining and petrochemical feedstocks and burner tip and internal combustion fuels.

Starting at the middle of the value chain, shown in figure 1–1, presents the easiest way to learn about the individual NGLs without getting into a lot of chemistry or physics, which will happen later. The more familiar names are there.

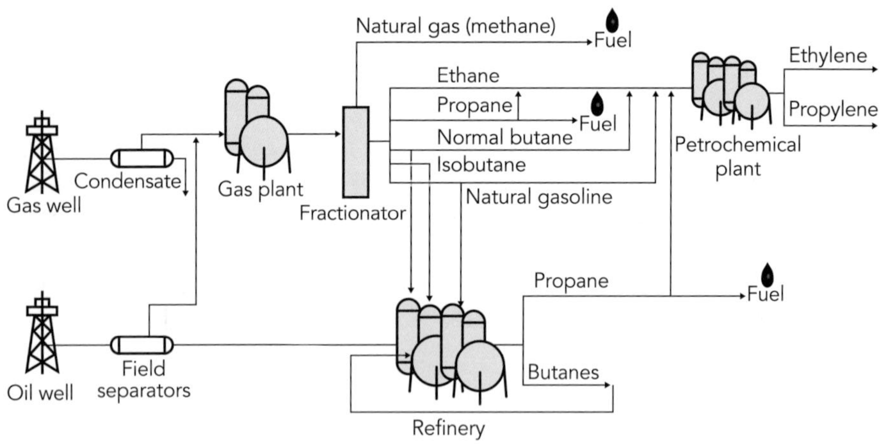

Fig. 1–1. NGLs value chain

Markets

Ethane, the simplest, lightest, and most volatile NGL, is used almost exclusively as a petrochemical feedstock. In those plants, the chain goes from ethane to ethylene and then to polyethylene or ethyl alcohol or dozens of other ethylene-based petrochemicals.

Propane, a volatile product, is shipped and handled under pressure to keep it in a liquid state. Propane may be the most versatile of the NGLs. Too heavy or valuable to leave in natural gas and too volatile to put in gasoline, it finds applications as a fuel for heating, cooking, and sometimes lighting (lanterns); as a petrochemical feedstock to make both ethylene and propylene as co-products; and in some places as a motor fuel. Petrochemical companies use propylene to make polypropylene, of course, and isopropyl alcohol, and dozens of other products.

Butane, less volatile than propane but still shipped and handled under pressure as a liquid, comes in two types. Normal butane has applications in the fuels market as a heating and cooking fuel, though not nearly as much as propane. Much of the butane supply is used as a gasoline-blending component to add a little volatility. That facilitates engine starts on cold mornings when the rest of the gasoline is a little less ignitable. Normal butane also goes to the petrochemical industry, where it is cracked to make the co-products ethylene and propylene and also butylene and butadiene. The process and the plants are just a little more complex than the ones that crack ethane or propane. Butylene and butadiene end up in plastics and rubber products. Isobutane, a different chemical structure than normal butane, finds a home in refineries. It has a chemical reactivity and molecular structure suitable for making high-octane gasoline-blending components called alkylate.

Natural gasoline might look and behave somewhat like gasoline. It is a liquid at room temperature and pressure. But by itself, it does not have the quality needed in today's automotive engines. The octane number is too low; the sulfur content is usually too high. But refineries can clean up natural gasoline and even upgrade it and blend it with better components to make it suitable for sale at gas stations. Natural gasoline is also used as a diluent for very heavy crude oils like Canadian bitumen. It reduces the viscosity of the crude oil so it can be moved in a pipeline. In some countries and even in some parts of the United States, natural gasoline is sometimes called plant condensate (but not here.)

Refineries

Moving along the diagram in figure 1–1, propane and the butanes come also from refineries. Most of the processing units in refineries, as they transform crude oil into refined products, work to change the chemical structures of the molecules that make up crude oil. The heat and pressure in the process units inevitably generate by-products, including propane and the butanes. The volumes are large enough that most refineries have their own gas plants onsite to capture the propane and butanes. They sell the propane but use the butanes themselves.

Gas Wells

On the left side of figure 1–1 are the origins of the NGLs. The stream coming out of a gas well can have anywhere between 80% and 100% gaseous material. (That ignores the entrained water, either vapor or liquid.) So at the well site, the stream goes into a field separator, which is more or less just a big tank. From the top comes the gaseous material, so-called wet gas or rich gas. (That is where the NGLs are.) From the middle comes oily material called condensate. From the bottom of the separator comes water.

Condensate

Condensate has been a perennial source of confusion in the industry. As the name implies, something has been condensed—turned from vapor to liquid. That would be an oily material that can be best described as a very light crude oil. Most times it has nothing heavier than diesel fuel-like material in it. Some people have called this material natural gasoline. Most people say that is wrong, that natural gasoline comes from a gas plant. In the old days, some called it drip gasoline or dripolene. In any event, producing, trading, and using condensate is a big business worldwide, but in this book it does not come under the topic of NGLs.

Meanwhile, back at the field separator, the vapor stream coming from the top is mostly methane. Depending on the well, it could have between as little as 2% or up to 20% NGLs in it. In most cases the stream goes to a gas plant for processing.

Oil Wells

Most oil wells have some gas and NGLs dissolved in the crude oil coming out of the ground. Some have almost 50%, more than can be dissolved, called *associated gas*. As the fluids from the reservoir come thousands of feet up the well and reach the surface, the pressure will have dropped considerably. Close to the well, the combined streams go to a field separator. Due to the much lower pressure in there, the gas and NGLs together spring from the oil. From there, they can go to a gas plant just as the gas well gas does.

Gas Plants

Gas plants perform two important functions, treatment and separation. Since natural gas coming out of the ground has contaminants with it (hydrogen sulfide, water vapor, carbon dioxide, etc.), the *conditioning* part of the plant takes care of those. In the separation part of the gas plant, what is generally thought of as natural gas, mostly methane is separated from some or almost all the NGLs, so the outturn of a gas plant is a stream of mixed NGLs and one of natural gas.

Fractionators

Fractionators separate NGLs from each other and make them available to the various markets. Fractionators can be located at the gas plant, or the mixture of the NGLs can move some distance, up to hundreds of miles away, to a fractionator. The mixture moving in a pipeline is called Y-grade (sometimes *raw make*).

But Wait! There's More

The definition of NGLs should be clear enough now—ethane, propane, the two butanes, and natural gasoline. But other, similar-sounding names and abbreviations come out of the oil patch.

LPG (liquefied petroleum gas). In the United States, LPG means propane meeting market specifications. In many parts of the world, LPG includes both propane and butane. Butane can be used separately as LPG as long as the temperature stays above 32°F. Below that, butane may not vaporize sufficiently to get to the burner tip or the engine as a gas.

Some people differentiate LPG from NGLs by their origin: LPG from refineries, and NGLs from gas plants. But a customer buying LPG (propane) for heating a house or for fueling an outdoor grill really does not care where the propane originated, whether a refinery or a gas plant.

LNG (liquefied natural gas). This is a completely different commodity. LNG is natural gas, 95%–98% methane, refrigerated to –165°F to turn it from vapor to liquid. With a reduction in volume by a factor of 600, LNG can then be transported across the sea by tanker or can be stored locally in steel tanks (still chilled to –165°F) for use during short periods of very high local natural gas demand (peak shaving). To get the LNG into a natural gas pipeline distribution system, it has to be turned back into vapor by heating.

A few other terms are worth mentioning. When you read the annual report of an oil and gas company and get past the glossy pages of write-ups and color pictures, you'll find the financials in the back, usually printed on dull gray paper. Notable on the first few lines are the production numbers for crude and condensate.

- Crude oil. Every type of crude oil is a unique mixture of light oils (such as gasoline), medium oils (kerosene, jet fuel, diesel fuel, or home heating oil), and heavy oils (industrial fuel oils and asphalt.) The portions determine whether to call it light, medium, or heavy crude oil. The weight (or density) of crude oil is measured by °API (American Petroleum Institute gravity, called degrees API) as follows:
    - Light crude oil has as API gravity higher than 31.1°API.
    - Medium oil has an API gravity between 22.3 and 31.1°API.
    - Heavy crude oil has an API gravity below 22.3°API.
    - Extra heavy oil has an API gravity below 10.0°API.

  Refineries, of course, separate crude oil into different fractions, process them, and then create finished products.

- Water. Water has an API gravity of 10.0°API.

- Condensate. As mentioned, condensate comes from the separator at a gas well. The properties from different gas fields vary considerably, but condensate is basically a very light crude oil, having little or no heavy oil in it. The buyers and traders take a close look at the condensate assay, a quantitative description of the properties, to understand whether the material is suitable for their use.

Onward

Thus the journey from the reservoir to market has quite a few entities grabbing their share of value improvement and usefulness as the NGLs make their way to market: producers, pipelines, processors, refiners, petrochemical plants, and marketers. The next 10 chapters cover them all.

CHAPTER 2

A Short History of the NGL Business

Beginning is easy. Continuing is hard.

Japanese proverb

"The available supply of gasoline, as is well known, is quite limited, and it behooves the farseeing men of the motor car industry to look for likely substitutes" (article in *The Horseless Age,* December 13, 1905).

"The world's total endowment of oil amounts to 60 billion barrels" (U.S. Geologic Survey, May 1920).

"The Stone Age did not end for lack of stone, and the Oil Age will end long before the world runs out of oil" (Sheikh Zaki Yamani, Oil Minister of Saudi Arabia during the 1970s).

"The USGS estimates that the mean undiscovered, conventional reserve additions in the world total 665 billion barrels of oil, 1,429 trillion cubic feet of natural gas, and 16 billion barrels of natural gas liquids" (U.S. Geologic Survey, June 2012).

Colonel Edwin Drake kicked off the oil and gas industry when he drilled his well in Titusville, Pennsylvania, in 1859. For the first 35 years the demand for lamp oil (kerosene) buoyed the growth of oil. But by 1890 people began to use light bulbs instead of lanterns, and oil's future turned uncertain. Then came the automobile. Up to that time, all that white stuff—gasoline and natural gas liquids—was considered a waste product. Most producers left the "wild" material, sometimes called fizz gas, greased oil, wet gas, or overhead reflux, in storage pits or open tanks to allow it to evaporate. Environmental concern was not high on the agenda in those days.

Early Market Development

By 1910 over 469,000 motor vehicles rode the American roads. The number doubled every two years for a while. Scores of automobile manufacturers proliferated (fig. 2–1). The demand for motor gasoline had grown way past the lamp oil demands. The rudimentary car engines were not particularly choosy about fuel. Oil refineries distilled a fraction of gasoline from crude oil that ran, well, well enough.

Fig. 2–1. Advertisement for a 1910 Ford Model T

The birth of the natural gas liquid (NGL) industry probably happened in 1910. Natural gas producers noticed that some of the condensate that came along with processing their product seemed to work reasonably well in the cars of the day. In their condensate plants, they compressed natural gas to pipeline pressures. Any oil in the gas plus a little propane and butane and almost all of what is now called a natural gasoline dropped out of the liquid as the pressures increased.

In the United States, German chemist Herman Blau patented a process for capturing some of the volatile gases, propane and butane, but it also included some heavier, oily materials in the gasoline range as well. The patent also covered storing the liquids in metal bottles. Nine different companies started up in the United States to sell Blaugas along with the expensive apparatus necessary to vaporize the gas, especially the heavy ends. Blaugas turned out to be a technical success and a commercial failure. The limited customer enthusiasm for the front-end investment just for cooking and lighting led quickly to the venture's collapse.

In 1910 Dr. Walter Snelling, then a scientist at the U.S. Geological Survey, built his prototype still, separated propane and butane from condensate, collected it in metal cylinders, and earned the everlasting title, "the father of propane" (fig. 2–2).

Fig. 2–2. Dr. Walter Snelling. His early work earned him the title, "the father of propane."

That same year, cousins Arthur and Chester Kerr had already built 20 plants around western Pennsylvania to make natural gas suitable for sale. The facilities included compression facilities that condensed the oily constituents of the natural gas, giving a volatile product sometimes called condensate, natural gasoline, drip gas, or even dripolene.

An interview reported in the Butane–Propane News in 1962 relates Arthur Kerr's experience. "The gasoline obtained from these plans was an ornery, volatile product. I tried to pass them (propane and butane) onto the gasoline trade but they boiled out of the liquid whether it was in storage, in transit, or in use and splashed red ink over our ledger. A daily loss of 1,400 gallons at that time amounted to $150, and no Scotsman could overlook that."

Kerr set about to separate out the propane and butane. He built a still that resembles something that a bootleg whiskey maker might have used, with coils, coolers, and kettles. He managed to separate the natural gasoline from the butane/propane mix, compress the volatile gases into a liquid, and store them in metal cylinders. With some marketing effort, he sold them to residential users for cooking and lighting.

Kerr's staff members were more shade-tree mechanics than engineers, and that led to more than a few incidents. In one case, a high-pressure liquid butane line burst, spewing ice cold vapor and freezing an operator to death. (This may actually have been the first of a series. Since then, notorious and inadvertent severing of buried liquefied petroleum gas (LPG) pipelines by excavation operators has had the same result.)

An interview by an *Indianapolis Star* reporter in 1912 tried to explain Snelling's technique:

> The natural gas is compressed and cooled and the heavy fractions which condense are separated. The lighter fractions are next condensed and are forced under pressure into a vessel called a "rectifier," where they come in contact with coils of superheated steam and are completely vaporized. The gas is then passed through a series of coils, and is heated [sic] to a lower temperature than the previous one, and these coils separate the gas into a series of products . . . to which the names "ethane," "propane," and "butane" have been given.

As sophisticated as this sounds, Snelling made the rig from an old hot water heater and parts scrounged from the laboratory at the U.S. Geological Survey laboratory.

The reporter got it almost right, but regardless, Snelling was so excited about his process that he patented it and began looking for commercial opportunities. Around the same time, Frank P. Peterson patented another piece of technology, liquefying propane and butane by refrigeration. Crude as it was, he packed ice around the coils of the vapor on its way to metal cylinders.

In 1911 Snelling joined the Kerr brothers to found American Gasol Company. In 1912, they had their first real customer, John Gahring, who allowed them to install a system to light the lamps in his house and fuel a gas stove. Recognizing the potential, only two years later, E. W. DeBower, the founder of LaSalle, America's most successful home study program, bought American Gasol for $50,000.

Unhappy with this turn of events, Snelling moved to Oklahoma and opened his own business. He bought condensate, distilled off the propane and butane, put them in iron cylinders, and sold what would later be called bottled gas or LPG, mostly for residential cooking and water heaters. A few customers used it for heating to supplement coal- or wood-burning furnaces. Alas, unable to shake off his bureaucratic tendencies, Snelling's business soon failed. He sold his patent to a local wildcatter, Frank Phillips, and he returned to the Bureau of Mines. But the LPG business had been hatched.

Arthur Kerr had not given up. In 1925 he moved to California and devised a simple mechanical system for use in rural heating and cooking. Success was boosted by assistance from Shell Oil Company, which quickly introduced their propane–propylene gas, Shellane, and Standard Oil Company of California (later Chevron), which marketed propane as Flamo. Soon Sun Oil Company introduced Solgas, propane recovered from its refining operations.

Calamity

Not much later, disaster in the drip gas business actually invigorated LPG. Few people had been paying enough attention to the hazards of drip gas and its dissolved propane and butane content. After all, it worked well enough in car engines. On a hot, 1915 summer afternoon in Ardmore, Oklahoma, a parked tank car of drip gas heated up enough to pop the pressure relief valve, whistling and spewing volatile gases. In a hapless attempt to deal with the issue, a railroad worker opened the fill valves of the tank to relieve the pressure, trying to avoid having the tank car explode. On that still day, propane and butane vapors saturated the adjoining area at ground level. Soon they reached an ignition source and an explosion leveled the town, killing 47 and injuring 500 people.

Gasoline makers quickly condemned the use of drip gas for motor fuel. By 1921 natural gas producers organized the Association of Natural Gasoline Manufacturers. (By that time, they smartly had purged the term *drip gasoline* from their lexicon.) They created new specifications for natural gasoline that limited the volatility. That forced all the unstable natural gasoline out of the automobile fuel market, but into the market came the culprits, propane and butane, sold again in cylinders for heating, cooking, lighting, and even blowtorches.

The development of novel ways to capture these increasingly popular gases, propane and butane, set the engineers to work. By applying the refrigeration principles of Frank Peterson, NGL recovery increased rapidly in the middle part of the 20th century. That fortuitously coincided with the extension of natural gas pipelines into the Midwest, where cold weather led to condensation of the NGLs, leaving slugs of liquid in the pipes (fig. 2–3). The introduction of lean oil technology first made ethane a marketable commodity, just in time for the nascent petrochemicals market to begin looking for cheap feedstock. Toward the latter part of the century, cryogenic turboexpander technology arrived just in time to feed ethylene plant growth and to keep natural gas pipelines that reached into the Northeast free of condensed liquids.

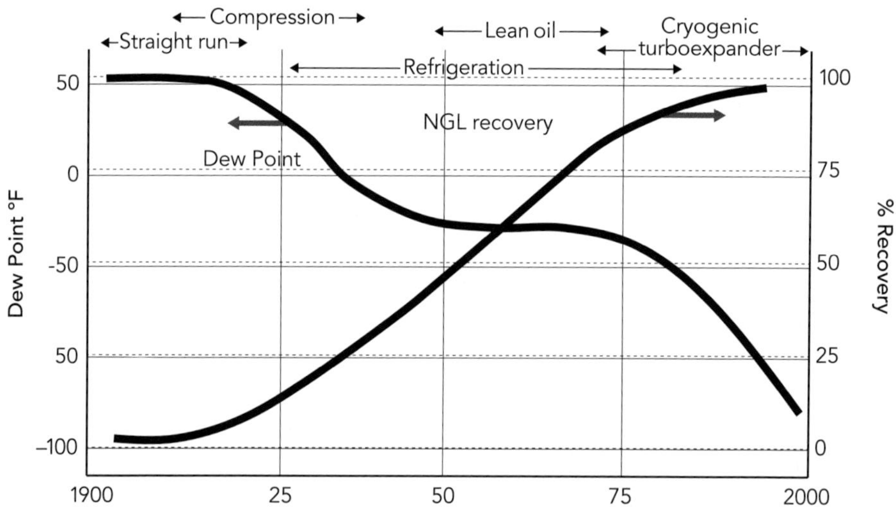

Fig. 2–3. The impact of technology on NGL recovery and dew points. New gas plant designs aided both NGL sales opportunities and the ability of natural gas pipelines to control condensation in their pipes, as measured by the natural gas dew point.

During the 1920s propane and butane marketers began to penetrate the agricultural market. First, they convinced farmers to convert tractors from gasoline to propane and butane. They had to overcome technical hurdles like seal integrity, valves searing, varying compression ratios, and carburetor stability. Second, they promoted propane and butane in crop drying, which proved easier. The clean-burning feature and lower prices per gallon made them an attractive alternative to kerosene.

Town gas pipeline distribution systems were a legacy of the 19th century in scores of cities around the United States and in advanced countries as well. The advent of cheap butane pushed town gas, which was made from coal, out of the picture. By 1931, butane/air mixtures had replaced town gas in 123 U.S. cities.

At about the same time, a Union Carbide chemist, George Curme, while trying to make acetylene from ethane, inadvertently made ethylene, today's largest volume petrochemical. (Acetylene was being used primarily as industrial cutting torch fuel.) This unexpected outcome made so much ethylene (and so little acetylene) that he turned his attention to finding uses for this opportune molecule. His first accomplishment was another blunder, making ethylene glycol while trying to make mustard gas, the poison gas being used in World War I. Based on all this serendipity, Union Carbide built an ethane cracker near the natural gas fields of West Virginia in 1920 to make ethylene in large quantities. That started the modern petrochemical industry. Shortly after came processes that produced polyvinyl acetate, ethyl alcohol (to the chagrin of contemporary prohibitionists), epoxies, polyethylene, and polyurethane, as well as an increasing demand for ethane and propane as feedstocks for ethylene plants.

More Calamity

The designers of the Hindenburg zeppelin thought they had a clever idea for international travel and demonstrated it on passenger airship trips between South America and the United States. They filled the bladders inside the zeppelin with hydrogen for buoyancy and ran the engines on butane. The energy content of butane per pound beat most other heavier fuels, even though it had to be stored in iron cylinders. Well, everyone knows what happened to the Hindenburg in 1937. Whether the butane fuel or the hydrogen had anything to do with the fire remains a contentious issue, but travel by any butane-fueled zeppelins abruptly came to an end.

> Ironically, the Hindenburg disaster made way for the use of LPG as a domestic fuel. The collapse of the aviation fuel market left thousands of pressurized bottles dormant around the world. In Rio de Janeiro, for example, Ernest Igal bought 6,000 of them and promoted butane as a cooking fuel to replace wood. By 1950 his company, Ultragaz, one of the world's largest LPG distributors, had 70,000 customers.

In 1937 a tragedy occurred in New London, Texas, when a natural gas explosion in a primary school killed 295 students and teachers (fig. 2–4). No one had detected the odorless gas leak in the basement. Researchers at the Bureau of Mines immediately accelerated their efforts to develop an odorant that would make leaking gas easily detectable. They found that no single odorant was perfect. People with colds and others who could not smell at all did not notice certain odors. Phillips Petroleum and Standard oil of California had already started to add odorant to their natural gas and LPG, jumping in before the Bureau of Mines issued its choice. They found good results with ethyl mercaptan, something the *Guinness World Book of Records* later called the smelliest substance on earth. The Bureau of Mines Pamphlet 58 set it as the U.S. standard in 1932, recommending the odorant at 1 pound per 10,000 gallons.

Fig. 2–4. New London schoolhouse after the 1937 natural gas explosion

Until World War II, some LPG dealers marketed propane, some butane, and some blends of the two. During World War II, the U.S. government requisitioned as much isobutane as it could in an effort to support the manufacture of aviation gasoline. The isobutane was used in a refinery process called alkylation, which produced a high-octane gasoline-blending component. Similarly, they set aside normal butane for a process that converted it to butadiene, a feedstock for making synthetic rubber, a commodity also in high demand during the war.

At the same time, the government rationed motor gasoline, and so farmers turned increasingly to LPG to fuel their tractors. Later, increasing demand for butane for gasoline blending gradually phased most of the butane out of the U.S. LPG market by the 1950s, effectively making the terms *LPG* and *propane* interchangeable.

Developing the Logistics

The introduction of pressurized LPG railcars in the 1930s spawned depots and bottling plant networks nationwide. Still, even as late as the 1950s, LPG retailers delivered their product in 100-pound cylinders, often off the back of pickup trucks. In 1960, Phillips 66's subsidiary, Philgas, built the first LPG bobtail, a medium-sized truck, with a large pressurized tank (fig. 2–5). Together with 500- to 5,000-gallon storage tanks placed at residential, commercial, and industrial sites, it allowed longer, more reliable reach and increased sales, with markets expanding from New England to the Pacific coast states.

Fig. 2–5. A vintage propane bobtail. This truck from about 1950 delivered both bottled gas and bulk propane.

In the late 1950s seamless pipeline technology had developed well enough that propane lines started reaching to distant markets in the U.S.–Mid-America pipeline to Minnesota, the Texas Eastern pipeline to New York, and the Dixie pipeline to Georgia—and points along the way. Ethane pipelines connected petrochemical plants to underground storage. Mixed gas liquids traveled by pipeline from Midland, Texas, and Cushing, Oklahoma, to Mont Belvieu, Texas, to be fractionated.

In 1927 three NGL businessmen, two from Phillips and one from Suburban Propane had organized the National Bottled Gas Association. By the choice of name, their foresight was somewhat limited. Eleven years later, more visionary members voted to change the name to the National Liquefied Petroleum Gas Association (NLPGA). However, it was not until 1987 that the bewilderingly lethargic NLPGA got around to changing its name to the National Propane Gas Association (NPGA). That late change occurred despite the fact that almost all butane had long been diverted from LPG, leaving only propane, and evidence such as the Chicago Transit Authority in 1950 ordering 1,000 propane-fueled buses and Milwaukee converting their 270 buses from gasoline to propane.

Underground storage

Expanding propane markets always presented another problem—the demand peaked in the winter and troughed in the summer months, but production remained flat year-round. Along came the flamboyant Garrison Haynes "Smokey" Billue, a sometime cowboy and gas plant operator. Around 1950 he noticed that many wells drilled in the in the West Texas area encountered thick layers of salt on the way to target. He reasoned that a cavity leached out of the salt could handle large volumes of propane in the summer, awaiting the winter demand. He convinced a few friends and possibly Phillips 66 to underwrite the first storage cavern, a 30,000-barrel cavity in the West Texas salt. Unfortunately for Smokey, he had not applied for or received a patent for his groundbreaking achievement. Over the following decades uncountable others have leached out millions of barrels of salt cavern storage, particularly in the great salt domes along the U.S. Gulf Coast.

Starting in the middle of the 20th century, major oil companies began withdrawing from the retail propane business. Shell got out in the 1950s, and Mobil in the 1960s. Multistate LPG marketers—Calgas, Empire Gas, Petrolane, and National Propane—grew by absorbing hundreds of small, local distributors. Ironically, only a few of these larger distributors survived in the long run, having been absorbed by other bigger enterprises like Amerigas and Ferrellgas.

Still, at the production end, literally thousands of producers proliferated in the oil patch. The largest gas producers built gas plants and pipelines from the prolific basins around the country and offshore—where the gas was—as well as internationally. But by the end of the century, many of these assets had been sold off to independent midstream companies that could bring the strategic focus and agility that large oil and gas companies could not or would not muster for this business sector.

International Markets and Trade

For the most part, NGL markets outside the United States developed more slowly, many 30 to 50 years later, even in the industrialized world (fig. 2–6).

Until World War II, the NGL industry remained a series of multilocal markets, a handful of operations around the world minimally connected by transportation. (The first purpose-built LPG ship was commissioned by Shell, built in Holland in 1931, and plied the European coast.) Europe started to move the fastest, not because of gas liquids coming from natural gas processing, but because refiners were looking for a home for their own propane and butane. Many refining companies moved downstream into bottled gas distribution to dispose of the by-product LPG.

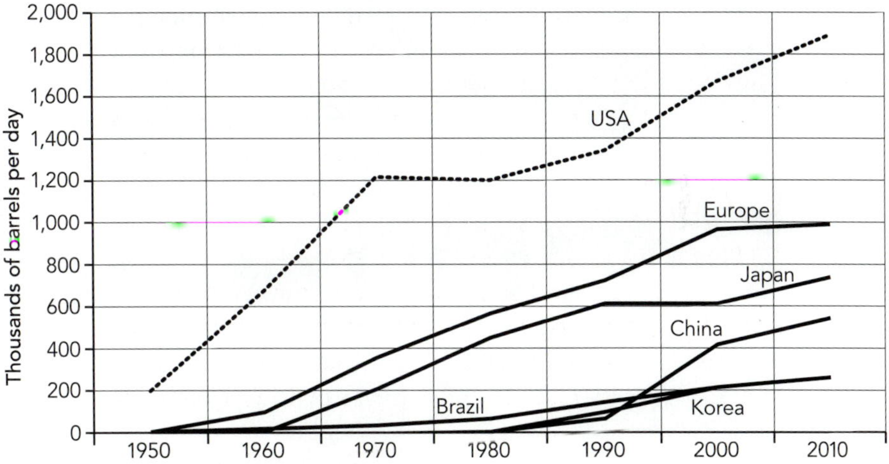

Fig. 2–6. LPG consumption by world area

After World War II, European refining companies began to increase their capacities as fuel oil replaced coal and as the number of gasoline- and diesel-powered cars grew. By the 1960s, coastwise LPG movement around Europe in pressurized tanks in small vessels connected markets. Northern European markets preferred their LPG as all propane, the more volatile gas. In the more moderate south, butane and propane/butane mixes were acceptable.

Marine transportation

The disparate markets around the rest of world had begun to be connected starting in about 1947. Warren Petroleum commissioned an LPG tanker, the *Natalie Warren*, (fig. 2–7) and began moving product from the Esso refinery in Aruba to Brazil and from the Houston ship channel to Newark. They had refitted a dry cargo freighter by installing 68 pressure vessels in five cargo holds to handle 100,000 gallons. Esso quickly followed with a small fleet of their own to ply the southern hemisphere.

Fig. 2–7. The *Natalie Warren*. The profile of this ship was sleek, but the installation of the 68 tanks on board was a plumber's nightmare.

The design of the pressure tanks was 240 psi. The tanks themselves weighed as much as the cargo, limiting the payload. To enable economic transport over long distances, ship designers had increased the payload. They had to reduce the thickness of the steel to contain it. The solution would be refrigeration: Cool down the liquid to reduce the volatility and the pressure needed to contain the vapors that came off the liquid. They designed three essentials:

- Onboard refrigeration equipment to maintain the cargo at 0°F
- Steel that was ductile enough to handle the lower temperatures
- Tank insulation to reduce heat transfer from the sea to the cargo

The first round of ships, dubbed *semirefrigerated,* dealt with all three.

The growing availability of LPG from the Persian Gulf found ready markets in Europe by the 1970s. Along with increased crude oil production came associated natural gas from the same wells. NGLs were part of the mix. Before 1970 most of the natural gas associated with crude oil production was flared. Bothered by conscience and by international abhorrence at the waste, the new national oil companies in the Persian Gulf states began to reinject the gas into the reservoirs for future use. Before doing that, they built natural gas processing plants to recover the NGLs for marketing.

A large shift in international trade patterns started around 1975. Most of the huge increases in Persian Gulf NGL production went to Japan, Korea, and China, as demand in those markets started to accelerate (fig. 2–6) with rising incomes, infrastructure development, and, importantly, the advent of larger LPG ships.

In 1962 the Mitsubishi shipyard delivered a new class of LPG ship, *the Bridgestone Maru,* known then as a VLGC (very large gas carrier). Refrigeration plants on this class of huge vessels kept the cargo at –45°F, below propane boiling temperature. That allowed virtually zero working pressure, with less tank weight (but more expensive nickel-steel alloy). The tanks no longer had to be cylindrical and could match the contours of the ship's hull. Some VLGCs now carry more than 20 million gallons of LPG (fig. 2–8).

Fig. 2–8. VLGC—very large gas carrier for LPG

Bad Forecasts and Bad Laws

In the 1970s and 1980s, the U.S. outlook for natural gas and NGLs turned sour. Most U.S. forecasters thought that the resource base, the gas in place in the ground, would not support even a flat level of demand. Much of this expectation was based on the theory of a crusty old Shell Oil Company geologist named M. King Hubbert. In 1956 he had proposed his classic Hubbert curve, based on examining the rise and fall of production curves of numerous oil and gas fields. He extrapolated that to countries, continents, and the world (fig. 2–9).

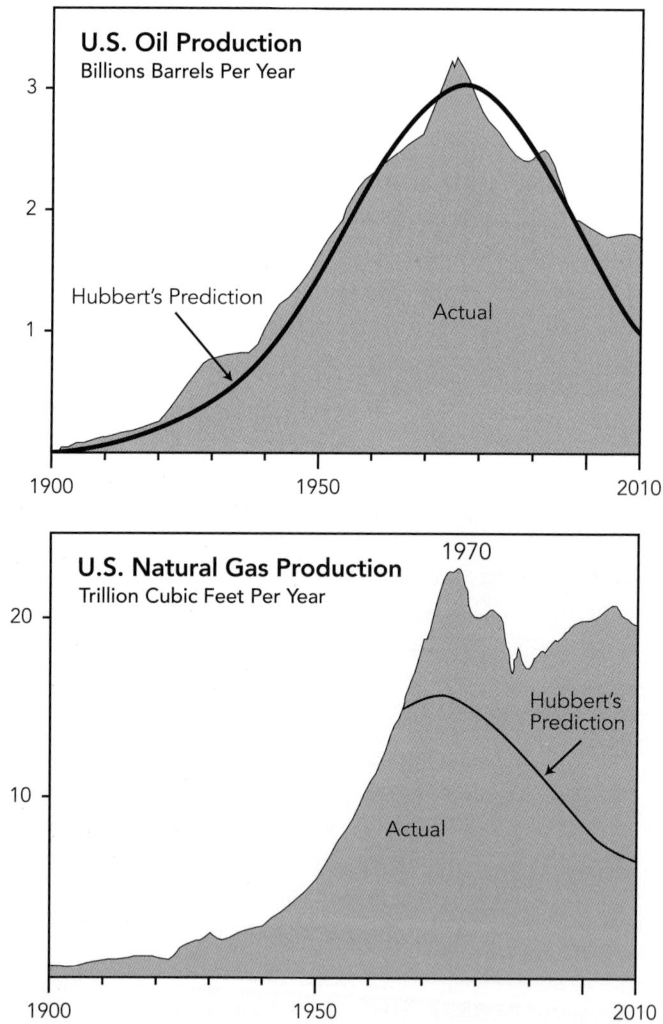

Fig. 2–9. The Hubbert curves. Hubbert's predictions in 1956 influenced outlooks for many decades.

The theory was convincing, especially when U.S. oil production did indeed start to decline in 1970, as he had predicted. As for natural gas, the decline also started around 1970, but after it had exceeded Hubbert's curve for a decade. Many forecasters at the time rationalized that fact by blaming logistical bottlenecks plus government controls that kept natural gas prices and production too low in the 1950s, but opened both up in the 1970s. In any event, along with the anticipated natural gas production decline, NGLs' availability was expected to wither.

Companies began to plan and build terminals to import liquefied natural gas (LNG), from the Middle East, West Africa, and the Caribbean. Petrochemical companies began to build ethylene plants in the United States designed to run on naphtha instead of ethane or propane. At the same time they built ethane/propane crackers in the Middle East where they thought the cheap feedstocks there would last indefinitely.

Compounding this general feeling of commercial pessimism was the ongoing U.S. regulation of the natural gas industry. U.S. price controls on natural gas were blamed for shortages that abruptly became critical about 1970. Federal regulators curtailed natural gas supplies to discretionary users.

Desperate to meet the demands of natural gas customers, the pipeline companies sought novel ways to boost their natural gas supplies. They built a few plants that converted oil and NGLs into a natural gas supplement called synthetic natural gas (another oxymoron, of course). Other companies installed facilities that would pump a mixture of propane and air into the natural gas pipelines. The combination approximated the content of natural gas. The propane, of course, had been previously removed from the natural gas at gas plants.

These ludicrous schemes worked because of the absence of arbitrage that an open market would have quickly shut down. The U.S. government had been regulating *interstate* natural gas prices on a cobbled-up cost basis for decades. The price of natural gas in unregulated *intrastate* business exceeded federally controlled interstate service by a factor of 4.

As the shortages became more acute in the 1970s, natural gas pipeline companies could buy (propane) high and sell (interstate natural gas) low. They still made money by rolling the propane cost into their too-low-priced natural gas, adding their allowable margin to the controlled natural gas price, which was still under what the market would bear. And they increased their volume at the same time.

In 1973 the Organization of the Petroleum Exporting Countries (OPEC) instigated an oil supply crisis. The U.S. federal government instituted price controls on all crude oil and products, including propane. More ludicrous schemes to game the system proliferated. Propane prices from gas plants were kept lower than propane from refineries since they were based on low natural gas prices, not high-priced crude oil. Propane traders were known to sit around a certain restaurant table in Houston in the mid-1970s, passing invoices around the table. Starting with a parcel of gas plant propane, each trader would buy and then immediately sell to the trader in the next seat, adding on an allowable margin until the final price reached the refinery-sourced level. This "daisy-chaining" lasted until President Reagan's administration dismantled the price and supply controls in 1983, effectively undoing all the ridiculously artificial schemes that government regulations created. By then, some propane traders were retired millionaires.

Turnaround

The pessimistic outlook for natural gas and NGLs from the 1970s lingered on through the 1990s and even past the millennium. But, a few entrepreneurial businessmen and engineers were not convinced. George Mitchell, like many others, knew that large volumes of *tight gas* lay in shale formations that abounded worldwide. No one had yet figured out how to extract it economically. Mitchell had the engineers of his company apply two novel technologies—horizontal drilling and hydraulic fracturing (fracking)—to these granite-like formations. That released large volumes of both light crude oil in some and natural gas in others. By the 2000s, other companies had started exploiting Mitchell's insights by cracking open, literally, the enormous resources of oil and NGL-rich gas locked in shale. By 2010, shale gas and the rich content of NGLs that came with it completely turned attitudes, outlooks, incentives, and opportunities upside down for producers, midstream companies, marketers, and petrochemical companies in the United States and around the world—anywhere shale oil and gas lay waiting. Production of oil, gas, and NGLs began to increase; petrochemical plant construction came back to the United States; and new pipelines and pipeline reversals abounded. Such a great quantity of NGLs came out of shale gas exploitation that NGL exports, especially propane, from terminals along the U.S. Gulf Coast increased rapidly.

Takeaways from a Short History of the NGL Business

- At the beginning of the 20th century, a few entrepreneurs, such as Snelling, Peterson, and the Kerrs, kick-started the NGL industry with their inventions that captured wild gas and made it marketable. At the time, all interest focused on drip gas, a volatile natural gasoline used for motor fuel, and propane/butane mixes for heating, lighting, and cooking.

- By the middle of the century, natural gasoline and most of the butane had been redirected as refinery feed; ethane and propane became interesting feedstocks for the nascent petrochemicals industry; and distributors delivered propane to millions of households.
- After the gloom of the latter part of the 20th century surrounding natural gas resources in the United States and the future of the NGL industry, the burgeoning supplies from shale gas led to surpluses and increasing U.S. exports.

Exercises

1. How did industry react to calamities?
2. What were some breakthrough technologies that impacted the development of NGLs?
3. What turned multilocal NGL markets into global markets?
4. How have major wars influenced NGLs?

CHAPTER 3

The Science and Engineering of the Upstream

One man's "magic" is another man's engineering.

—Robert Heinlein (1907–1988), *Time Enough for Love*

In 1859, Colonel Edwin Drake sought his fortune in western Pennsylvania, having watched the price of lamp oil made from whale blubber escalate intolerably. He had heard of oil seeps in the hills around Titusville and hired a water well man, now known only by the name Uncle Billy, to dig (not drill) a well near one of the seeps. Uncle Billy's crude rig pounded a hole in the ground with a chisel at the end of a shaft, lifted and dropped by cable. Periodically, he had to pull the chisel from the hole and remove the rubble. At 69½ feet, oil began to bubble to the surface. Billy sent his gofer into town to tell Colonel Drake the good tidings: the modern oil age had just begun.

This may seem like really deep background for some, but for anyone interested in the origins of natural gas liquids (NGLs), here is a light treatment of petroleum geology, geophysics, and petroleum engineering affecting oil, gas, and NGL creation and production.

Origins

The story that excites geologists starts with the Big Bang theory. Whatever or whoever caused it to happen, 14 billion years ago or so, untold numbers of particles—mostly hydrogen, the simplest atom that existed in primeval clouds of gas—swirled in space. By the forces of gravity, they concentrated into fireballs or stars. Some became stable, and some collapsed on themselves in spectacular convulsions, sending more debris into space. Some hydrogen atoms morphed into helium and later into carbon, nitrogen, iron, nickel, and other elements.

Of most interest to the subject of NGLs is the eventual collection of enough material into solar systems and more importantly into the origins of the planet Earth. As gravity forced heavier particles to the center of this swirling soup, lighter elements floated to the surface. At the end of this unimaginably slow evolution emerged Earth, a big sphere with a solid center, an inner core, and a molten layer around the center (the outer core). Outside that, a deep layer called the mantle has remained surprisingly viscous because of its proximity to the heat of the molten outer core. Finally, on top of that layer lies the crust upon which people eventually appeared (fig. 3–1).

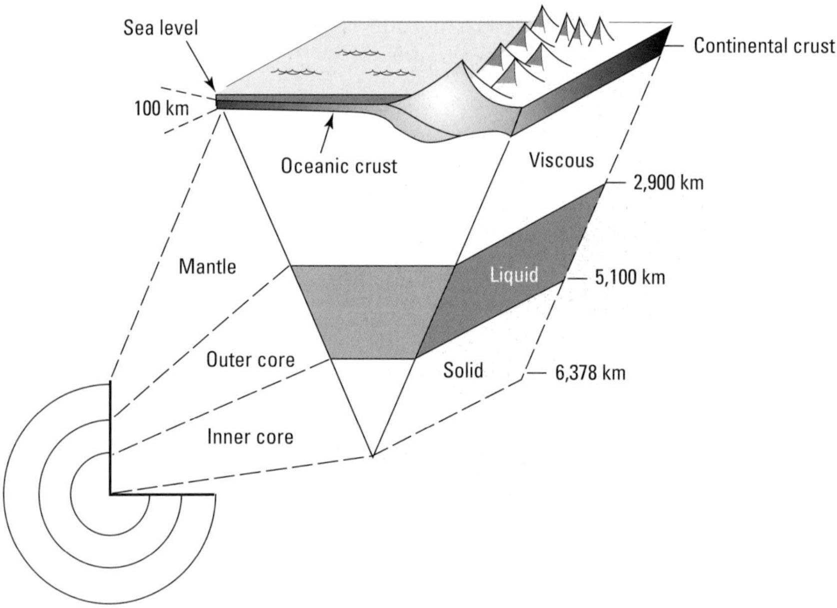

Fig. 3–1. Earth's layers

The scientists who think about these things concluded that enough heavenly rains landed on the surface to cover most of the planet with water. Still, a mammoth landmass now called Pangaea stuck out. (Pangaea comes from a Greek word meaning "the whole Earth.") As the planet cooled over a few billion years, up until only about 225 million years ago, rising molten rock sundered this large piece of acreage into numerous plates. Since then, the plates have moved around the surface of the Earth, floating on that viscous mantle like lily pads on a pond (albeit at a smaller rate) in a phenomenon called plate tectonics. Eventually, as they wandered about the planet, they came to the positions and shapes we know today as the continents and the seven seas.

As tectonic plates moved about, they bumped and pushed against each other, sometimes causing great deformations in the topography. We see them now as mountains and valleys. Sometimes one plate slipped under another, causing great uplift on the upper one. Regardless of how they got there, mountains endured continuous abuse from wind, rain, freezing, and thawing. Erosion sent big and small pieces of rubble down into streams and rivers, washing them into lakes and oceans (fig. 3–2).

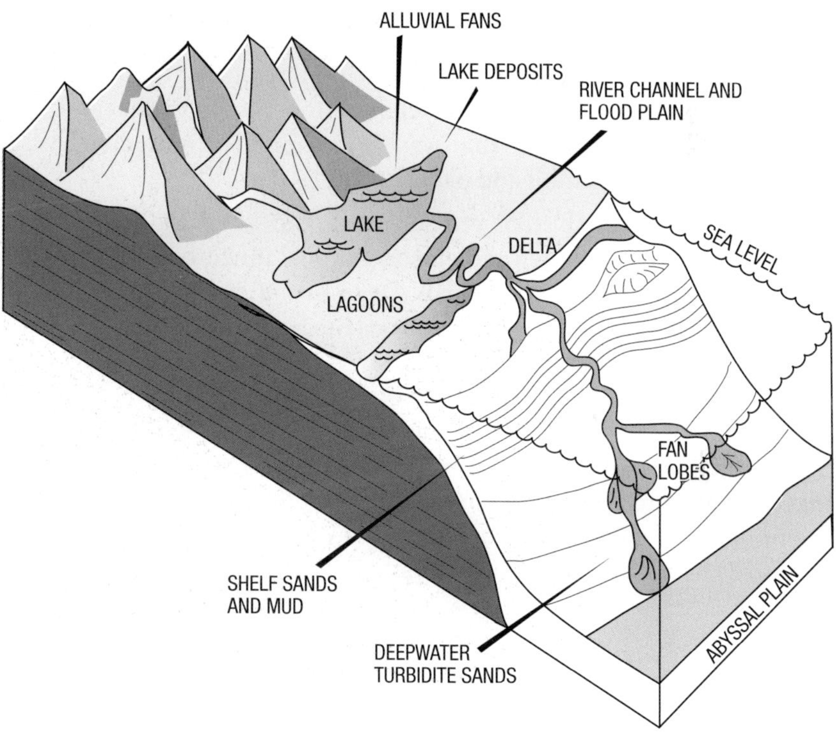

Fig. 3–2. Erosion of Earth's surface

Enter Microorganisms

Meanwhile, as long as 3.5 billion years ago, early life-forms—microorganisms—appeared and apparently made their way into the more hospitable parts of the planet, onshore and at sea. For a few billion years, as the debris coursed down the waterways, sometimes copious amounts of microorganisms traveled along with it, coming to rest in ancient lakes, river beds, and oceans. More of them thrived there. Subsequently, millions of years' worth of sediment washed down, covering them. Eventually, thousands of feet of rock, gravel, sand, and clay, had covered up many of these organic rich accumulations. Old bodies of water, like the one that covered much of Texas, disappeared after years of sedimentation, leaving rich accumulations of organic material buried well beneath the dry surface.

Cooking Up Hydrocarbons

All that sedimentation brings the story closer to its climax. In some places, those of the most interest to petroleum geologists, the microorganisms sat in zones conducive enough to convert them into oil and gas. Those places were under high pressure and temperature—high pressure because of the thousands of feet of sediment above; high temperatures because of their depth, ever closer to that molten layer near the Earth's center and to the hot viscous mantle around it. That was the recipe—the ingredient list and cooking instructions—for oil and gas creation.

In the next, essential step, enough heat and pressure to "cook" the microorganisms changed their molecular composition into oil or gas or both. The heat came from inside the Earth. Deep in the planet's innards is that layer of hot, molten material that surrounds the Earth's solid center core. The temperature of everything outside that layer—ever closer to the surface—gradually decreases. So the deeper from the Earth's surface, the hotter the subsurface gets. Having the ingredients (the microorganisms) hot enough, for long enough, often yielded oil and gas. However, too little heat sometimes changed nothing; too much cooked it all the way to graphite. Scientists can create in the lab a small-scale version of this chemical metamorphosis, but the massive change by natural forces took millions of years.

Migration

After this improbable chemical transformation of living matter, these hydrocarbons sometimes migrated (fig. 3–3). Hydrocarbons are lighter than the water that saturates much of the Earth's subsurface down for miles, and they do not mix well. So by gravity, oil and gas could rise toward the surface. If they found an impervious barrier, called a trap, and a comfortable place to accumulate, called a reservoir, they came to rest. Traps and reservoir rock come in many combinations, determined by the geologic forces that created the various subsurface characteristics, including sedimentation history, tectonic plate movement, settling, and erosion. And any or all of those could have caused faults or pathways in the subsurface from the source rocks.

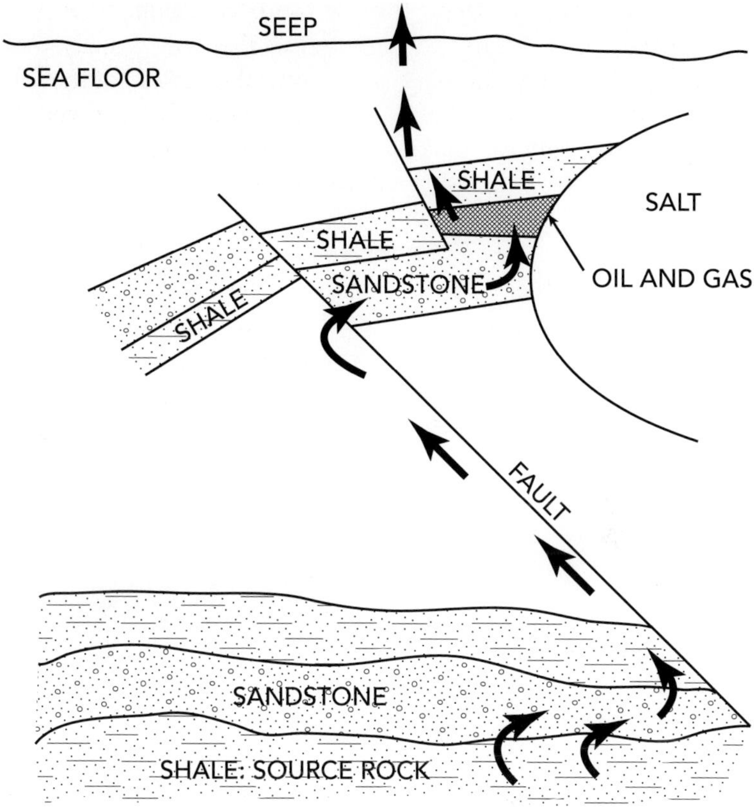

Fig. 3–3. Migration of oil and gas

Sometimes oil and gas migrated all the way to the surface, which is why ancient societies found oil pools on the ground that they could harvest and burn for lighting and use as medicine and sealants for their canoes and ships. Chinese records show that people piped seeping gas to palaces for lighting as early as 200 AD. That was the clue that brought Colonel Drake to Titusville, Pennsylvania, in 1859.

Reservoir Characteristics

Despite casual perceptions, rock, especially reservoir rock, is not solid. Much of it is made up of fine grains that are packed together by overburden pressure (fig. 3–4). Oil and gas (and NGLs) sit in the spaces among the grains. The granular structure and the pressure in the reservoir play a large part in determining how productive the reservoir might be (how it will flow) when a well penetrates it and opens access to the surface.

Fig. 3–4. Reservoir rock magnified 50 times by an electron microscope

Shale Oil and Gas

The hydrocarbons residing in shale, which has very tight granular packing, might have formed right there in the shale or migrated there over a very long period of time. In either case, producing them effectively requires more than just drilling a well into them. The granular structure is so fine that it inhibits fluid flow. The shale has to be fractured to open up channels for the oil and gas to move to the well.

Fracking technology uses water pumped at high pressures into horizontal wells that run through the beds of shale to create a fine lattice of fractures that create pathways for the oil and gas. Completing a hydraulic fracturing (fracking) job requires dozens of pieces of equipment at the surface to inject the water at pressures sometimes exceeding 50,000 psi (fig. 3–5). A special kind of sand, called proppant, is pumped along with the water. The proppant remains in the fractures after the pressure is released and keeps passageways propped open to allow the oil and gas to flow to the wellbore and up to the surface.

Fig. 3–5. Equipment assembled for a fracking job

In some shale gas accumulations, the percent of NGLs, especially ethane, is unusually high. That makes processing the gas to remove the NGLs an imperative to meet pipeline specifications for maximum heat content, or dew point control, and for Wobbe index compliance. (See chapter 4 and appendices 1 and 2 for details.)

Summing up, the essential four ingredients for exploitable accumulations of oil and gas (fig. 3–2) are the following:

- Source rocks, where the transformation took place
- Migration paths
- Reservoirs, sediment often called sands, where the hydrocarbon accumulated
- Traps, the impervious layer of rock or maybe salt layers, that sealed off the reservoir

Geologists and Geophysicists

Colonel Edwin Drake was a practical man but neither an engineer nor a scientist. Upon his oil discovery in 1859, he immediately rigged an old water well pump to lift the oil into a wash basin. He sent a man into town to buy up all the barrels he could find—pickle, cracker, or otherwise. That spread the news about his find, and shortly a hoard of oilmen raced to Pennsylvania and then into Ohio with scores of theories about where the richest oil reservoirs lay: under hills, along their sides, along northeast-southwest trends, or parallel to river beds. It took 40 years before the scientific discipline began to permeate the industry as petroleum exploration companies hired their first bona fide geologists.

Many petroleum geologists now spend their careers studying and mapping the subsurface in search of the right combination of the essential four: source rock, migration, reservoir, and trap. They examine the geological history. What was the result of the tectonic plate movements? How did the various layers of sediment accumulate? Where did the sediment come from, and what was the composition? Was the organic matter in place deeply enough and long enough to undergo transformation? Are there suitable reservoir rocks and traps?

Geologists team up with geophysicists, who use their own tools, particularly seismic technology, to gather information about the structure and composition of the subsurface. Today, seismic survey crews lay parallel trails, maybe a mile long, of geophones along the ground. (Offshore, seismic vessels tow mile-long strings of hydrophones.) The crew sets off a large detonation or otherwise generates sound waves, and the receivers pick up the echoes. As the sound bounces off each succeeding, deeper layer of sediment deposited in ancient times (fig. 3–6), each echo (signal) is recorded. Trillions, quadrillions, or more data points are captured. Huge-capacity computers later make sense of all these signals, using complex mathematical algorithms. Work on these subsurface techniques and algorithms continues today.

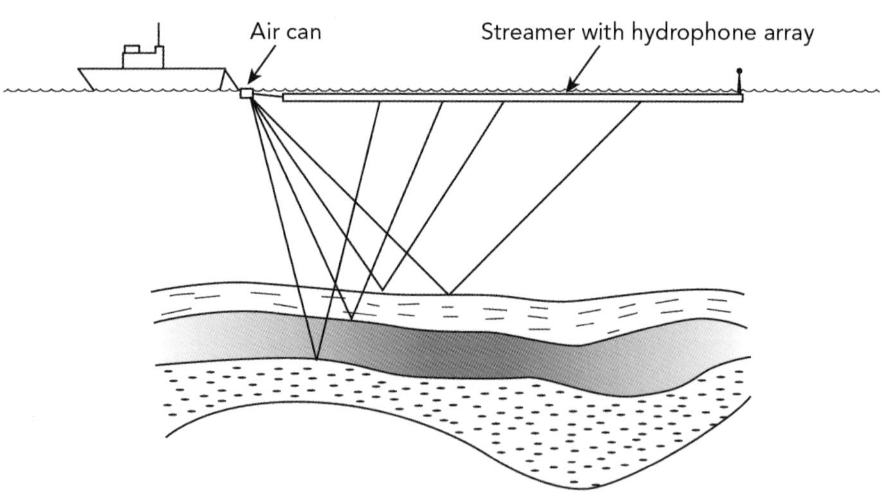

Fig. 3–6. Seismic vessels generating and capturing (recording) subsurface signals

In the early years of seismic development, geophysicists made line drawings of just a slice of the subsurface. Sometimes they hand-colored the lines to highlight the features—layers, faults, and other geologic anomalies, plus potential reservoirs full of hydrocarbons. Toward the end of the 20th century, computing power had increased sufficiently to process and display the subsurface in brilliantly colored three-dimensional projections. Reservoirs with a high probability of hydrocarbon accumulations became more apparent. Life for explorers became a little bit easier as they used these tools to help them hypothesize about the existence of oil and gas deposits. But there's nothing like penetrating an oil- and gas-bearing reservoir with a drill bit to prove a hypothesis.

Drilling

After the geologists have formed a hypothesis that a hydrocarbon-laden reservoir exists, the engineers take over. After they both get approval from whoever is going to put up the funds, the engineers have to design a well that will safely reach target depth. They will take into account what they know about the subsurface on the way to the target depth. Information sources can include seismic data and nearby wells. They have to know the target depth, the route that the well must take (vertical, slanted, horizontal), the type of rock in the various layers that are to be penetrated, subsurface pressures from gas or water to be encountered, and more.

If they are going to spud the well onshore, they will have to secure a land rig right-sized for the depth. The deeper the targeted depth, the larger the rig has to be. A 20,000-foot well needs a derrick that can lift more than 350,000 pounds of tubulars (drill pipe, casing, production tubing). For an offshore well, a jack-up or a floating drilling rig has to be hired. In the deepwater, the derrick may have more than a million pounds of lifting capacity. Drilling onshore or offshore has many of the same steps, but offshore the drill bit may have to travel down a few hundred to maybe 10,000 feet or more through seawater before making hole at the ocean's floor.

Spudding a well

To start, the driller (using the rig, of course) spuds a hole where a conductor, a stubby pipe, is cemented in place. This acts as a guide for the drill pipe and a foundation for a blowout preventer (BOP). In the event of an uncontrolled surge in pressure in the well while drilling, the driller can seal off the well using the hydraulic rams in the BOP.

The driller inserts a drill bit at the end of the drill pipe through the BOP and the conductor and begins "turning to the right" to drill the well (fig. 3–7).

As the drill bit chews through the ground, the driller pumps drilling mud down the inside of the drill pipe. The mud goes through holes in the drill bit, keeping it from overheating as it drills. Then the mud returns up the borehole—outside the drill pipe but inside the hole, of course. The mud also carries the rock cuttings with it, keeping the bottom of the hole clear.

The mud also provides wellbore integrity. The deeper the hole, the more pressure comes from the surrounding formation (the layers of sediment outside the wellbore wall), because of all the overburden (the arrows in fig. 3–7). The heavy mud provides counterpressure so that the wellbore does not collapse. The driller ensures that the deeper the hole, the heavier the consistency of the mud.

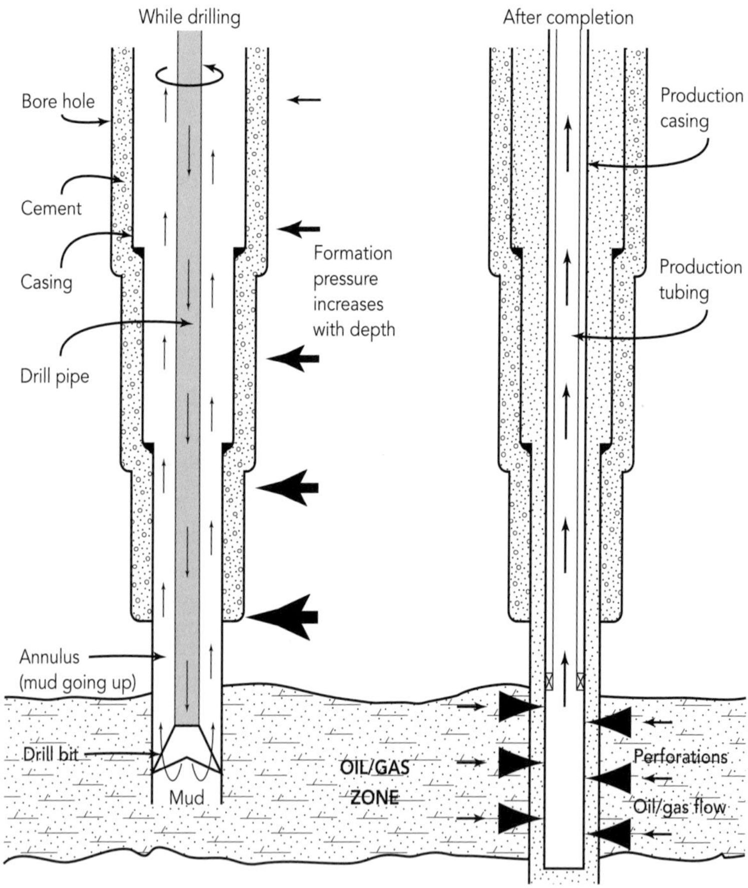

Fig. 3–7. Drilling and completion. On the left, multiple lengths of ever-decreasing diameter casing have been set, and drilling mud flows down the drill pipe and up the annulus around it. On the right, the cemented-in production casing has been perforated in the oil/gas zone to allow flow to the surface through the production tubing.

Setting casing

Eventually, the mud gets so heavy that it might start to move into the surrounding layers of sediment at the upper part of the well hole. At that point, the driller stops drilling and pulls the drill pipe and bit out. This seemingly archaic and arduous effort, called *making a trip*, takes hours of effort by the rig crew. (They have to do this whenever they have to change a worn-out drill bit as well.) They have to pull the drill pipe out, length by length, wedge it so that the rest does not fall back in the hole, unscrew it, stack it, and pull the next length out.

Once the hole is clear of the drill pipe, the driller inserts steel casing into the hole, again adding it, piece by piece, to the bottom of the hole. The driller then pumps cement down through the casing and up its outside, between the casing and the wall of the wellbore. The cement and casing provide a permanent barrier, ensuring the integrity of the well and preventing disastrous collapse. At the same time, the steel casing prevents any fluids from the wellbore from penetrating the walls of the well and contaminating aquifers, water tables, or any other formations. Bedsides that environmental imperative, loss of mud into the formation would be no small money matter either.

After the drill bit and drill pipe go back in the hole (piece by piece) though the casing, drilling resumes. As the bit goes deeper, the driller has to repeat the process every time the drilling mud gets too heavy for the formation below the last casing string. Of course, to fit in the hole, every additional set of casing has to be a smaller diameter than the one above it. The drill bit has to be smaller as well. The driller has to carefully balance mud and casing in wells in order to reach depths of 25,000 feet or more.

While the bit is turning in the hole, the driller watches the weight of the drill pipe pressing down on the drill bit. After a while, if the driller were to allow the full weight of the drill string to sit in the bottom, it may well crush the drill bit. While drilling, the driller continually receives measurements of pressure, temperature, formation composition, and fluid content—usually water, but maybe some hydrocarbon along the way. Geologists examine the cuttings that came up with the mud. They look for formation characteristics and the presence of water or hydrocarbons.

Finally, the driller reaches the target depth. Several things can happen at this point:

- If nothing of commercial interest has been found, the driller can abandon the well by pulling all the tubulars and plugging the hole with cement.
- The driller can run a short drillstem test, letting the fluids in the reservoir flow thorough the drill pipe.

- Electronic logging of the well can take place to better define what has been found.
- The well can be temporarily abandoned, pending further information from the cuttings, nearby wells, seismic surveys, or analysis.
- The well can be completed.

Logging

Like a fortune teller reading tea leaves, the geologist has been examining the cuttings as the drill bit went ever deeper, as well as the log the driller keeps. That gives a beginning picture of what is down there. When the target depth has been reached, logging a well can significantly improve that image. Operators use many different types of logging techniques. They all operate in a similar mechanical way. A *sonde,* a signal transmitter/receiver of some kind of signal —electrical, nuclear, gamma rays, and others—is dropped by wire down the well (fig. 3–8). A recording device captures the signaled data as the sonde passes through the formations. Depending on the type of log, the mapped signals indicate the nature of the formation, the type of fluid it contains (water, oil, gas), and the location (depth) and thickness of the reservoirs.

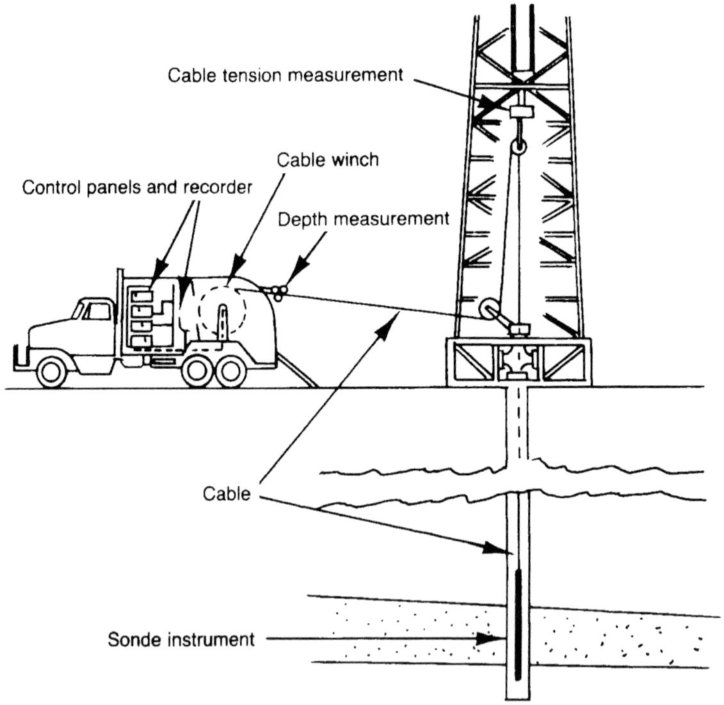

Fig. 3–8. Logging a well

Completion

In a happy case, the logs, and maybe the drillstem test, show the well to have commercial quantities of hydrocarbons and sufficient flow to warrant completion. The driller runs yet another set of tubulars, production casing, down the well and cements it place. The driller inserts production tubing inside the production casing, with some packers at the bottom. (See the right-hand side of fig. 3–7.) That provides a path for the oil or gas to flow or be pumped to the surface through the production tubing.

Finally, in the "pay zone," where the logs have shown that the reservoir sands and the hydrocarbons lie, the driller perforates the production casing by shooting holes in it, literally, with a perforating gun. Natural gas wells and many oil wells have sufficient pressure in the reservoir to push the reservoir fluids—oil, gas, and some (or a lot of) water—to the surface. A set of valves, called a Christmas tree, replaces the BOP to control the flow at the surface.

For low-pressure oil wells that are onshore, the iconic pump jack at the surface operates the pump at the bottom of the well to provide the needed boost. Offshore, where space on a platform is precious, an electric pump at the bottom of the well might do it. In some deepwater wells, an electric pump at the sea floor gives enough boost to get the fluids up the thousands of feet to the production platform.

Caution

All this makes drilling a well as easy as boiling water without burning it. Not really. A list as long as a length of drill pipe worries drillers as they penetrate the subsurface:

- Lost circulation—losing mud into the formation
- Breaking a pipe in the hole and having to fish it out
- Keeping the casing centered while cementing it
- Drill bit wandering off vertical
- Turning the bit for horizontal drilling
- Measuring vertical depth and trajectory to the target
- Production tubing perforations plugged with sand or gravel
- And on and on

Surface Operations

The well is completed, and the fluids are ready to flow. For an oil and gas field of any size, the petroleum engineers have to devise a development plan. Reservoir engineers study the characteristics of the reservoir (size, produceability measured by the reservoir rock characteristics, composition of the contained fluids, pressure). They determine how many additional wells need to be drilled and their spacing in order to best exploit what has been found. They determine where the water injection wells can be placed best to dispose of water separated from the produced fluids.

Offshore fields add another level of complexity. The type of production platform will depend on the depth of the water, weather patterns, pipeline access to onshore, and more.

Separation

During production operations at an oil well, as the fluids reach the surface, they enter a field separator (fig. 3–9), a vessel not much more than a horizontal or vertical tank. The pressure and temperature will have dropped between the reservoir and the surface. It drops even further as the fluids enter the voluminous space in the separator. At that reduced pressure and temperature, the vaporous natural gas and NGLs separate from the oil. They rise and exit at the top of the separator. Oil and any water fall to the bottom and naturally separate, as oil and water do, with water exiting the lower valve. The middle layer of crude oil comes out of the side valve. The process at a gas well is more or less the same, except oily material dissolved in the gas condenses (which is why it is called condensate) and comes out of the middle of a gas well separator.

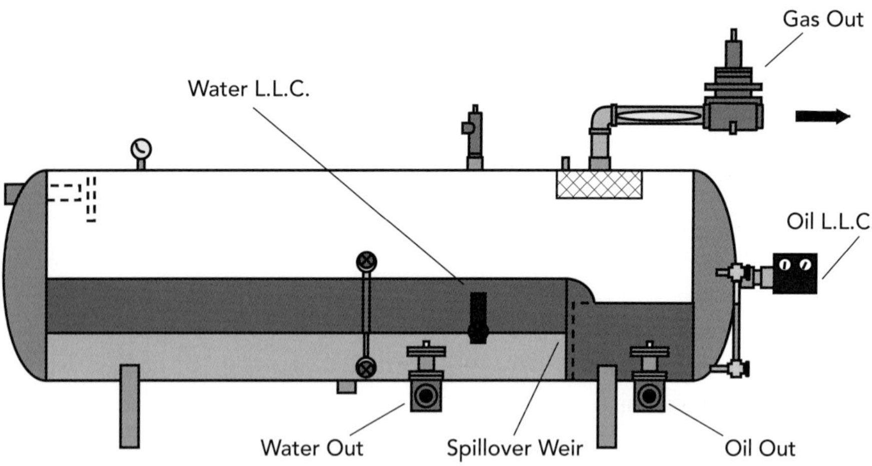

Fig. 3–9. Horizontal field separator. Production from the well separates under reduced pressure.

At a point just downstream of the separator, custody of the oil and gas changes. The natural gas with whatever NGLs are dissolved in it goes through a meter into a gas pipeline, probably on the way to a gas plant. The oil or the condensate goes to local tankage onshore and ultimately to a truck or an oil pipeline for transport to a refinery. Offshore, the oil goes into a small holding tank and then into a pipeline or tanker to move ashore. The water from the separator is usually reinjected into the reservoir, with two benefits. It solves a water-disposal issue, and it adds energy to the reservoir, boosting hydrocarbon recovery. The upstream enterprise has finished its work.

More caution

Do not think that the reservoir engineers and production engineers do not have their own issues as they keep the fluids flowing after the surface facilities are in place:

- Reservoir pressure control
- Rate of flow through the reservoir
- Undrained areas of the reservoir
- Fluid lifting techniques as pressure drops
- Reservoir water management
- Separating any emulsions of water and oil at the surface
- And more here also

All this makes for a good read in other PennWell books that dwell on these subjects . . . but not here.

Takeaways from the Science and Engineering of the Upstream[1]

- Geologists and geophysicists use geologic surveys and seismic technology in their search for the four essential factors that could deliver them a productive oil and gas well:
  - Source rocks
  - Migration
  - Reservoir rocks
  - Traps

1 If this terse summary goes in the opposite direction of your curiosity, much more detail is available in *Oil & Gas Production in Nontechnical Language*, by Martin S. Raymond and William L. Leffler. Tulsa OK: PennWell, 2006.

- Drillers use drilling rigs, drill pipe, and drilling bits to burrow into the Earth. They set casing along the way to protect the integrity of both the well and the subsurface.
- If the appraisal of the drilled well by logging and other analysis is positive, the well can go through the completion process that prepares it for safe production.
- At the surface of a producing oil well, engineers direct production to a field separator, where any natural gas and NGLs come out the top, crude oil comes out the middle, and water exits the bottom. From a separator for a natural gas well, natural gas and NGLs come out the top again; condensate, much like a very light crude oil, comes out the side; and water comes out the bottom.
- The natural gas and NGLs usually go to a gas plant for separation.

Exercises

1. What are the four geological essentials that result in a subsurface accumulation of oil and gas?
2. What makes shale oil and gas more difficult to produce than conventional oil and gas?
3. What kinds of tubular goods (pipes) have been put in the ground by the time a well is completed?
4. What happens when the fluids in a well reach the surface?

CHAPTER 4

The Chemistry and Physics of NGL

The only real science is physics. All the rest are just stamp collecting.

—E. C. Rutherford (1871–1937), physicist, who had to eat these words in 1908 when he accepted the Nobel Prize in chemistry

About Hafnium

When it comes to elements, chemists never had any love of or respect for physicists. Still, the physicists struggled continuously in their labs searching for elements missing from the periodic table of elements.

In 1922, Danish physicist Niels Bohr, with inscrutable insight, used simple geometry and quantum mechanics to conclude that the yet undiscovered element, number 72, should not be placed next to the next lighter element, lutetium, number 71. It could not hold any more electrons in its inner shells. It was a new breed. He dispatched two physicists to examine his idea, and on their first try, they discovered number 72 and named it hafnium, based on the Latin word for Copenhagen. It was, indeed, a quantum leap from lutetium.

After that, chemists slowly and begrudgingly began to accept that atomic structure, quantum mechanics, and physicists did have something to do with the chemical attributes of elements.

Just a little *chemistry* goes a long way for natural gas liquids (NGLs). It can demystify why propane is called C_3, but not C3's, and why butanes can be called C4's. Unfortunately, chemistry is mostly about memorizing . . . or remembering where in the book you read it.

The *physics* of NGLs appeals more to the senses. It has to do with temperatures pressures and even feel. They are so much more intuitive. Physical properties figure importantly into commercial transactions.

Chemical Properties

Adam's atoms

These days, everyone knows what elements are, with their familiar and sometimes bizarre names (gold, nitrogen, neon, tellurium, niobium, etc.). At this writing, 118 different elements have been identified, but scientists at the world-class labs at Livermore, California, at Darmstadt, Germany, and at Dubna, Russia, occasionally find others, to which they give names like einsteinium, copernicium, and livermorium, darmstadtium, and dubnium. Really. Atoms are the basic units of those elements, and combining different kinds of atoms together makes up molecules or compounds like ammonia, rust (iron oxide), nylon, and even skin.

Even though you can touch those things, the atoms and molecules themselves are unimaginably small . . . and versatile. Consider that when Adam first stood up and scratched himself, he was a Masterful assembly of an uncountable number of atoms, on the order of 6 billion billion billion of them. (A chemist would say 6×10^{27}.) And they all apparently worked well together, because here we are somewhat later, each happily activated with our own unique 6 billion billion billion.

Vital force

Like most things in life, when it comes to chemical compounds, there are two kinds, organic and inorganic. Up to the 19th century, chemists thought that organic chemicals, those made of carbon and hydrogen, came only from living matter—vegetable or animal or people. Hence the name organic. Then in 1828, German chemist Friedrich Wöller synthesized urea in his laboratory using inorganic compounds. (Why he was trying to make urea, a constituent of urine, is obscure.) That completely undermined the vital force theory of organic compounds so popular at the time. But chemists are not wont to change quickly, so today's classification, organic and inorganic compounds, remains the same:

- Organic compounds—molecules made up of carbon and hydrogen atoms (called carbons and hydrogens after this), and sometimes a little oxygen, nitrogen, sulfur, and some other kinds of atoms. Depending on who is doing the calling, they are sometimes called by the names *hydrocarbons or petrochemicals.*
- Inorganic compounds—all the other kinds of molecules that do not contain both carbon and hydrogen.

Rules of molecules

What most people do not know about molecules is that chemists say that they have identified over 1 million different chemical compounds, but bewilderingly, over 90% of them are organic chemicals. Why would that be?

It has to do with the carbon atom and three simple rules for making molecules:

- Rule 1: *Valence.* Some atoms have a compulsion to attract themselves to other atoms. Carbon has a valence of 4. It wants to share its electrons promiscuously with four other atoms. Hydrogen monogamously has a valence of 1. It wants only a single partner. All other elements have their own valences: 0, 1, 2, and so forth.
- Rule 2: *Bonds.* In this book, and many others, bonds are shown as dashes. In reality, atoms bond with each other by sharing electrons using an arcane mechanism involving quantum physics. Hardly anyone understands quantum physics, so they revert to using dashes, which gets the job done for most.
- Rule 3. *Isomerization.* A combination of the same atoms may attach themselves to each other in several different structures without violating the rules of valences and bonds. (It is like building something else with the same set of Legos.)

Displaying and discussing the different molecular structures that the NGLs take on will help understand why the NGLs can be used in both different and similar ways.

Paraffins

Start with the simplest organic chemical, methane. As in figure 4–1, methane has one carbon and four hydrogens. That satisfies the valence rule—each carbon is attached to four other atoms, each hydrogen is attached to one other atom.

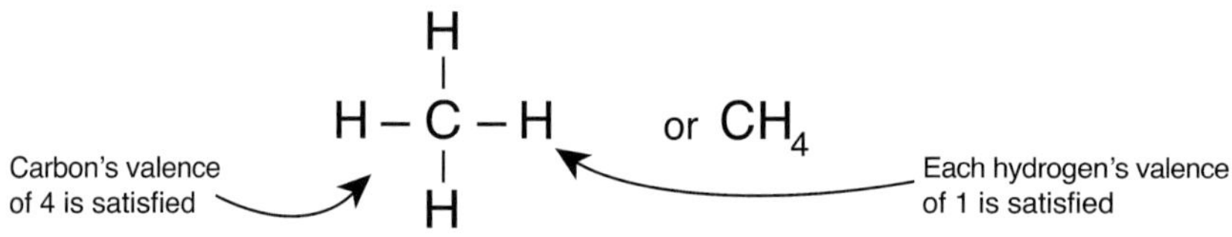

Fig. 4–1. The methane molecule

Ethane is only slightly different (fig. 4–2). Take a hydrogen off methane, leaving a *methyl group*, and replace it with another methyl group, one containing a carbon and three hydrogens. All the valence rules are then satisfied. Each carbon is connected to four other atoms; each hydrogen is connected to one another atom. That process can be extended beyond ethane to get propane, normal butane, and normal pentane. The chains can go up to 30 or 40 carbons or more (plus the accompanying hydrogens.) This is a family of hydrocarbons known as *paraffins.*

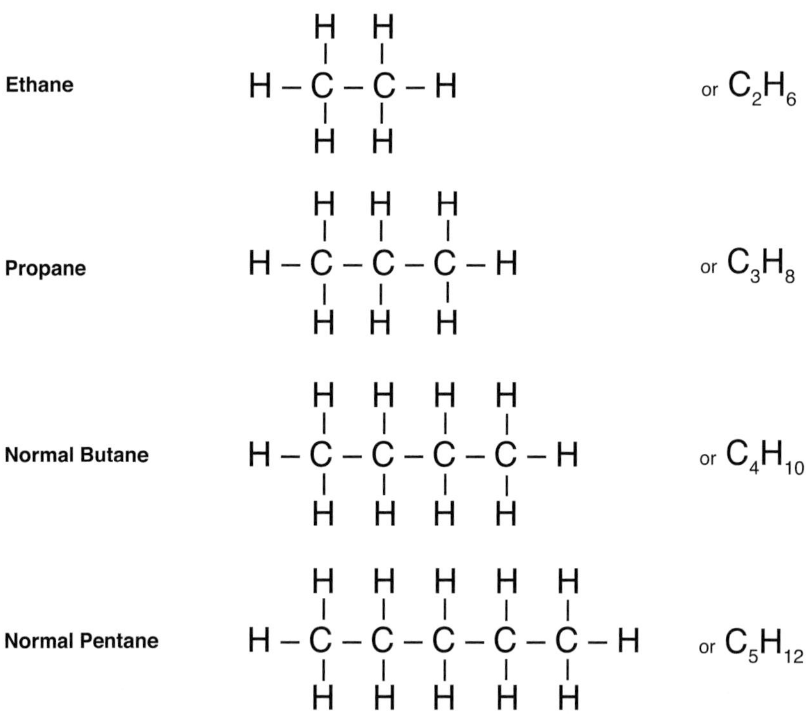

Fig. 4–2. Ethane, propane, normal butane, and normal pentane

Isomers

There are other ways of assembling these combinations of carbon and hydrogen atoms—the *isomers*. In figure 4–3, the isobutane molecule has the same carbon and hydrogen count as the normal butane molecule. The label "normal" distinguishes it from its isomer. Isobutane has a *branch* from one of the middle carbons made up of a methyl group, but still has all of the valences satisfied by the bonds . . . and the same carbon and hydrogen count. Same story goes for normal pentane and isopentane, and so on. For those increasingly larger molecules, the branches can be in different places, so the number of isomers increases rapidly with the number of carbons.

Normal butane
(straight chain)

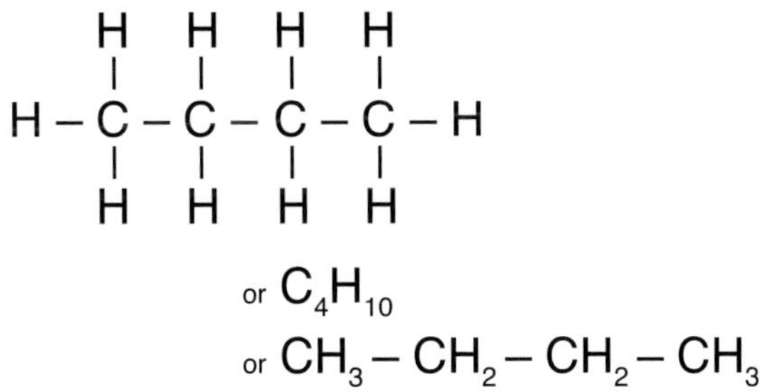

Isobutane
(branched chain)

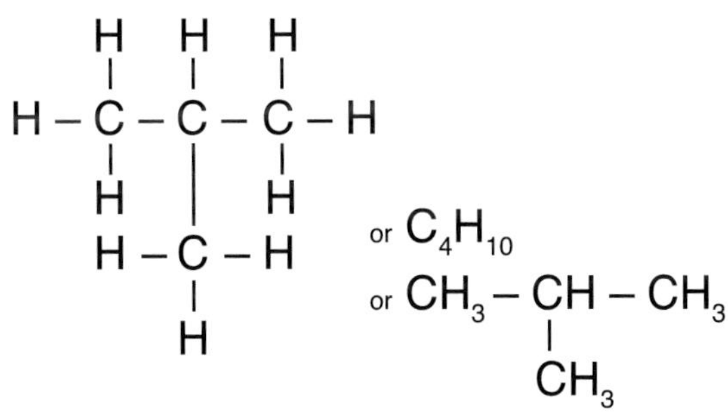

Fig. 4–3. Butane isomers. Both have the same number of carbon and hydrogen atoms, but different structures.

These isoparaffins are cousins or a subset of the paraffin family. As with any family members, they are different from each other. Even though both butanes, normal and iso-, have the formula C_4H_{10}, they each have unique properties—weight, volatility, chemical reactivity, boiling temperatures, and others. The same uniqueness goes for the isomers of C_5H_{12} or, say, a $C_{16}H_{34}$ molecule, and so on. The more carbons, the more isomers with their unique properties. The same isomer rule goes for the molecules in the next categories as well.

Olefins

In the chemical industry, molecules are often manipulated by breaking them up. That is the situation in an olefins plant. In cases like that, after breaking, say, an ethane molecule in two, the carbon atom might not be able to find enough hydrogen atoms around to satisfy its passionate need for four partners. Ethylene is the result and a prime example. It is "missing" some hydrogen (fig. 4–4). To deal with that, *two* bonds connect the carbons, satisfying the valence of carbon with four connections. (In reality, the carbons share four electrons, and chemists call that a *double bond*.)

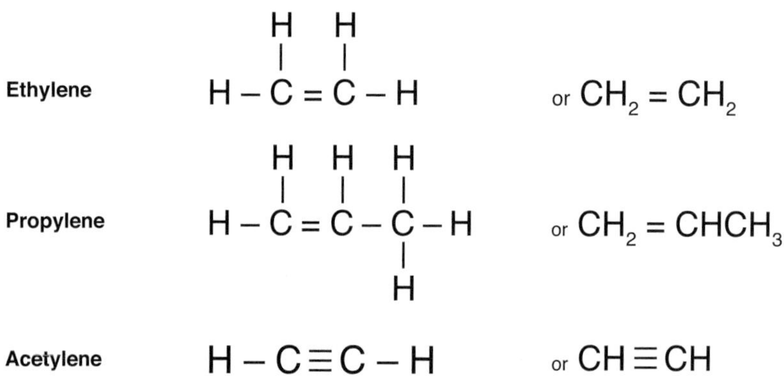

Fig. 4–4. Ethylene and propylene

Propylene has the same situation between two of its three carbons and is "missing" two hydrogens. Ethylene and propylene are the first two members of the *olefin* family. Both these molecules are very reactive: They have a strong urge to eliminate that double bond, replacing it with a single bond by attaching the carbons to other atoms or groups of atoms. That is why ethylene and propylene are the two biggest base chemicals or building blocks in the petrochemical industry. Scores of petrochemicals start with these two.

Olefins, by the way, are manufactured, not naturally occurring. They come from cracking operations in refineries or chemical plants. Thus olefins are not in the list of NGLs. But they have to be included here in order to grapple with chapter 8, on petrochemical demand for NGLs.

The larger olefins—butylene and beyond—also have double bonds and can also have branched structures just like the paraffins can, for example, isobutylene (C_4H_8) or, say, isooctene (C_8H_{16}), and so on. Olefins get created in the cracking processes in refineries, too, so oil products like gasoline and diesel can have plenty of olefins, though they are bigger molecules than ethylene or propylene. That is why people do not leave gasoline in their lawn mowers over the winter. Those olefins with double bonds will find ways to react with anything, including each other, forming gums and varnishes, making it aggravatingly difficult to start the mower in the spring.

More than most other people, chemists have their own nomenclature, and it turns up in the olefins. Instead of butylene, they say *butene*; occasionally, they use the terms *propene* for propylene and *ethene* for ethylene, especially if they are from Britain.

Cyclics

Not all organic chemicals (hydrocarbons) have to be in a straight or branched chain to satisfy the three rules. The common cyclic compounds in figure 4–5 are like straight chains with their ends connected. Because of that, they need fewer hydrogens than the paraffins do to satisfy the valences: C_6H_{12} instead of the straight chain C_6H_{14}. The cyclic compounds—cyclopentane and cyclohexane—are commonly found in natural gasoline. This third class comprises *cyclic* hydrocarbons, usually called *naphthenes* by refiners.

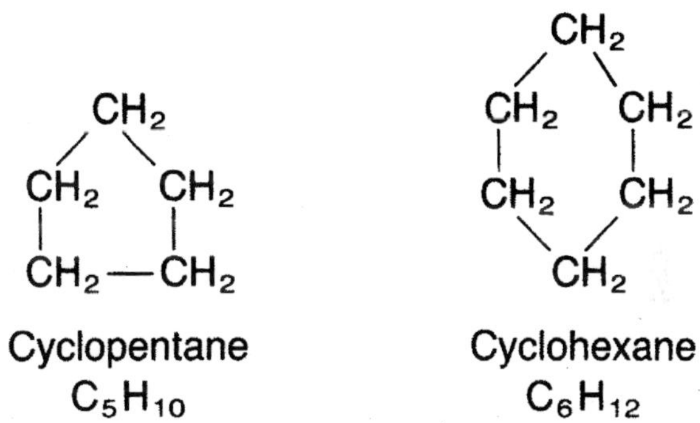

Fig. 4–5. Cyclic hydrocarbons

Aromatics

Another (and blessedly final) class of compounds covered in this book, the aromatics, seem to be like both the cyclics and the olefins. The three compounds in figure 4–6, benzene, toluene, and xylene, have cyclic structures but double bonds between every other carbon in the ring. That satisfies the valence rule, but it makes these three compounds very reactive and puts them in the pantheon of base petrochemicals, up there with ethylene and propylene.

Toluene and xylene have some notable features. Instead of a hydrogen attached to all the carbons, toluene has attached to it the *methyl group.* That, of course, gives toluene properties different from benzene. And, by the way, benzene and toluene have no isomers, but xylene has three. The second methyl group in xylene can attach to three different carbons around the ring, and that results in three isomers, each with distinct properties. One of them, called paraxylene, is the source of the ubiquitous PET (polyethylene terephthalate) water bottles. Numerous other petrochemical derivatives come from the aromatics, for example, styrene, polyurethane, and nylon.

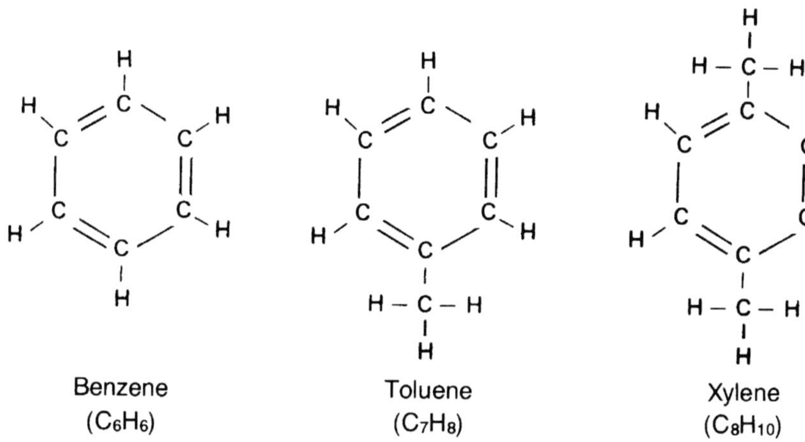

Fig. 4–6. The aromatics—benzene, toluene, and xylene

Combinations and natural gasoline

By combining some of the paraffins, cyclics, and aromatics (but not olefins), and taking into account isomers, scores of molecules can make up natural gasoline. The next sentence eases into the shorthand of organic chemicals by leaving off the hydrogens: Forget for a moment the simpler C_5 and C_6 paraffins. Forget the simple cyclic and aromatic hydrocarbons for the moment. The number of other unique structures increases staggeringly when seven or eight carbons are arranged in ways that have paraffins, cyclics, or aromatics *hooked to each other.* See the dimethyl cyclopentane and ethyl benzene in figure 4–7. All these molecules and many others can immeasurably proliferate natural gasoline.

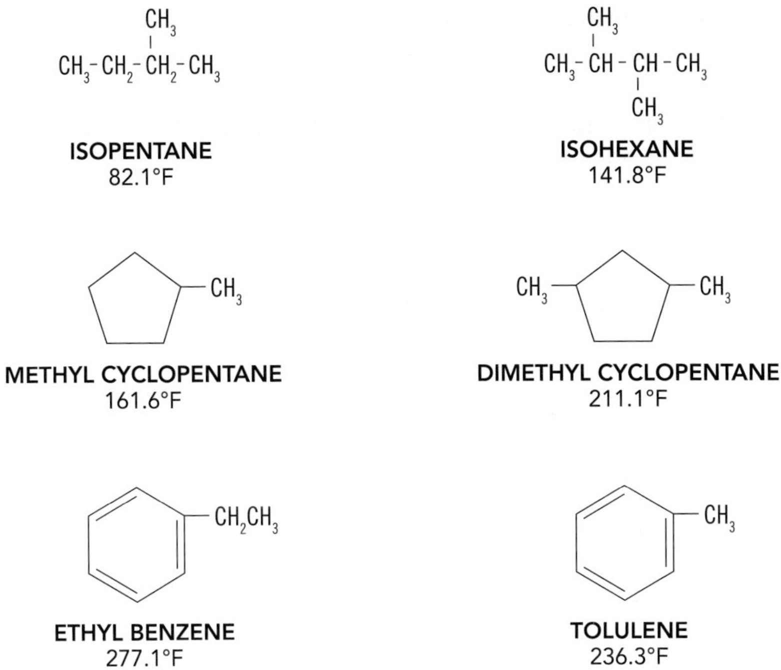

Fig. 4–7. Some of the many different molecules found in natural gasoline and their boiling points.

Physical Properties

The four NGLs—ethane, propane, the butanes, and natural gasoline—sometimes compete with each other in the same markets. In others, they have exclusive advantage. Much of that depends on their unique physical properties: volatility, octane number, energy content, boiling points, and so on.

Boiling temperatures

A crucial property of the gas liquids is called *boiling* temperature, which is the same as the *liquefying* (vapor-to-liquid) temperature, but easier to think about and spell. The boiling temperatures of the NGLs are shown in table 4–1. Some salient points come from these numbers.

- To make LNG (liquefied natural gas) for storage or transport, natural gas has to be cooled (an understatement) below –259°F.
- Ethane, propane, and the butanes are usually kept at pressures higher than atmospheric pressure, enough to liquefy them. (Higher pressure raises the boiling temperature.) That way, the NGL can be moved around as a liquid, a more easily transportable form.

- Some tankers and storage tanks use refrigeration reduce the pressure of propane and butane to reduce the amount of steel they have to use for the tanks. Others use more refrigeration to drop the temperature below the boiling point in order to store them as a liquid at or near ambient pressures.
- If you take a butane heater or lantern on a camping trip, you had better hope that the temperature does not drop below freezing, 32°F. If it does, normal butane will not vaporize—no gas will come out to the burner tip to ignite. (You will have to stick the butane canister under your coat for a while to heat up the fuel. After that, the burning fuel will provide enough ambient heat to keep it going.)
- Natural gasoline sometimes has some light ends—pentane or even butanes dissolved in it, so some vapors are likely to come off at ambient temperatures. That creates a hazardous condition if any flames are nearby. (The same is true for motor gasoline, which is why gas stations post "no smoking" signs.)

Table 4–1. Boiling points at atmospheric pressure

NGL	Boiling Point
Methane	−259.0°F
Ethane	−127.5°F
Propane	−43.8°F
Isobutane	10.90°F
Normal butane	31.1°F
Natural gasoline	Starts at ~50.0°F

Volatility

Some trivia about NGL volatility:

- Ethane has a very high volatility requiring high pressures to contain it, so it cannot be stored in handy sized cylinders for use in backyard grills.
- Propane's volatility is lower so it can be stored in this way.
- Normal butane's volatility is lower than propane's and lower than even isobutane's. It can be used in gasoline, even though it is a gas at room temperature and pressure.

Some of these properties are evident in table 4–2. Motor fuel volatility is measured by Reid vapor pressure (RVP). Vapor pressure measures how easily a liquid evaporates, that is, turns from liquid to vapor.

In order to minimize the amount of gasoline fumes getting into the atmosphere, governments regulate the allowable RVPs of motor gasoline. The limits are set typically in the 8.5–12 RVP range, depending on the season and the geographic location. More on this is found in chapter 9, on the fuel use of NGLs.

Table 4–2. Vapor pressures and octane numbers

	RVP	Octane Number (R + M)/2 Method*
Ethane (at 70°F)	543	—
Propane (at 100°F)	190	107
Normal butane	52	92
Isobutane	71	100
Natural gasoline	10–14	65–80
Motor gasoline—max	8–12	87–93

*This common measurement seen on gas station pumps; see appendix 4 for explanation.

Practices in the petroleum industry deal with these RVP properties. Reasonably obvious from the RVP column in table 4–2 are the following:

- Motor gasoline can tolerate *no* ethane or propane added to it.
- A greater amount of normal butane than isobutane can be added to gasoline because normal butane has a lower RVP and does not buck up against the motor gasoline RVP limit as fast as isobutane does.
- Blending a little natural gasoline into motor gasoline increases the vapor pressure just a little.

Propane warrants a little more special treatment here because it is such a widespread consumer product, found in backyard tanks piped into homes and in cylinders under patio grills. Vapor pressure makes those sources operable. As vapor leaves the container, more liquid *flashes*, that is, turns from liquid to vapor. That happens to be an *endothermic* process, which means that the propane absorbs heat from its surroundings, causing their temperature to drop, and so the propane and its container cool a little (or a lot, maybe forming frost on the container or its valves if the vaporization is really rapid such as in a bad leak). The same effect happens when someone holds a wet finger up in the air to find the wind direction. The side of the finger that cools (because the moisture is vaporizing, drawing heat from the finger) tells where the wind is coming from. Of course, nowadays golfers throw some grass in the air to do the same thing.

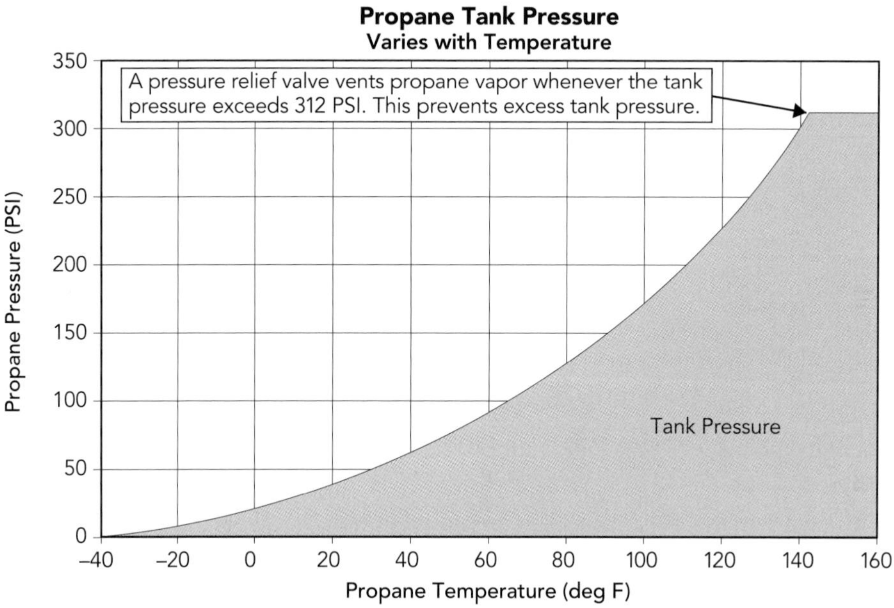

Fig. 4–8. Propane temperature versus vapor pressure. The tank is rated at 312 psig, and the pressure relief valve opens at that point.

The rate at which propane vaporizes is a function of temperature (fig. 4–8). The higher the temperature of the propane liquid, the higher is the pressure of the tank contents. Tanks in the United States used for stationary equipment such as irrigation pumps or generators have a maximum working pressure rating of either 250 psig (pounds per square inch read on the gauge) or 312 psig, depending on when they were manufactured. Stationary tanks such as those at residential and commercial locations have a rating of 250 psig. Propane cylinders operate between 240 and 312 psig. All these vessels have pressure relief valves designed to prevent the vessel from rupturing. The relief valves on cylinders pop off (open) at 375 psig; the other two types of tanks, at 250 or 312 psig. Figure 4–8 has a message for firefighters. If a propane tank is involved in a fire that is heating the tank, when the propane reaches the pop-off temperature, the vapor release can exacerbate the fire control problem. To understand how this is handled, see chapter 11.

Octane numbers

Most cars run on gasoline or diesel fuel, but propane has a small share in some markets. Internal combustion engines for gasoline or propane are designed to run on fuels with certain octane numbers. What does that mean? Octane numbers measure whether a certain fuel will preignite as the piston moves up in the cylinder, compressing (and heating up) the fuel/air mixture, and before the spark plug lights off. That causes inefficient and sometimes harmful operation of the engine. Bigger or more powerful cars have higher compression ratios—how long the piston travels in the cylinder. They need a fuel that is less likely to ignite as the piston compresses the gasoline/air mixture. Higher octane gasoline has less tendency to preignite.

Two different test methods measure octane number:

- The *research octane number* (RON) method simulates motoring along a highway
- The *motor octane number* (MON) method simulates going up a steep hill or pulling a large load like a boat.

In the United States the octane number posted on gasoline pumps is an average of the two measures and shows up on the pump as (R + M)/2. (See appendix 3 for a fuller explanation of octane numbers and their measurement.) Averaging those two numbers does not mean too much to most people, but that is the number mandated by the U.S. government regulation. Posted octane numbers vary across the country based on the local atmospheric pressures and temperatures. For example, in the Rocky Mountains the lowest grade might be 85(R + M)/2 for regular gasoline; in California it might be 87(R + M)/2.

All this endlessly tedious description is foreplay to understanding where natural gasoline fits in. The composition and therefore the octane number of natural gasoline produced from any field are unique. Some natural gasolines that are high in normal paraffin content can have octane numbers as low as 60(R + M)/2. Others high in aromatics could be in the 75 to 80 range.

Heating value details

1 Btu (British thermal unit) equals the amount of heat required to raise 1 pound of water by 1°F.

Americans and the British commonly use British thermal units in commerce. Scientists and much of the rest of the world use joules. Natural gas distribution companies bill their customers in therms.

1 Btu equals about 1,055 joules. One therm equals about 100,000 Btu.

Higher heating value (HHV) equals the heat, the number of Btu, given off by burning a particular fuel.

Lower heating value (LHV) equals the HHV less the amount needed to vaporize the H_2O created during combustion, which depends on the number of carbons and hydrogens in the particular type of molecule.

In either case, natural gasoline can be blended into gasoline at a refinery in limited percentages before it drags down the average octane number of the finished gasoline, which might have to be 85 to 93. Many refineries process parts of the natural gasoline to increase the octane number. For example, they may route some lighter parts of the natural gasoline to their *isomerization* units and some of the heavier parts to catalytic reforming. See chapter 6, on refineries, for more details.

Energy content

NGLs, like other hydrocarbons, burn. They give off heat when they go through an exothermic (heat generating) reaction with oxygen.

Burning methane: $CH_4 + 2O_2 \rightarrow CO_2 + 2H_2O$ + heat

Burning propane: $2C_3H_8 + 10O_2 \rightarrow 6CO_2 + 8H_2O$ + heat

On burning, each NGL gives up a different amount of heat, usually measured in British thermal units (Btu). Table 4–3 shows the heat content in terms of higher heating value (HHV) of NGLs and some other fuels that compete with them. (Another measure is the lower heating value (LHV), which takes into account the amount of energy necessary to vaporize the water that gets created in the process of burning.) The table makes it apparent that a stream of natural gas with a higher percentage of NGLs may exceed the requirement by pipeline companies for gas with a maximum HHV heat content of, in many cases, 1,100 Btu per standard cubic foot (Btu/SCF). For that reason (and others discussed in chapter 5) the gas plants remove enough NGLs to bring the energy content down.

Table 4–3. Energy content: Higher heating value

	Btu/SCF	Btu/Gal
Nitrogen and CO_2	0	—
Methane (CH_4)	1,010	—
Ethane (C_2H_6)	1,770	66,000
Propane (C_3H_8)	2,516	91,600
Normal butane (C_4H_{10})	3,369	102,916
Isobutane (C_4H_{10})	3,354	98,850
Natural gasoline	~5,000	~125,000
Kerosene	—	~134,000
Diesel fuel or home heating oil	—	~143,000

In a changing world, fuels compete with each other in many ways, including price, logistics, or as base load or peak shaving. A handy way to relate the values at various prices is the *peephole diagram* that first appeared in the *Oil and Gas Journal* in 1972, with updates in 1977 and 1980 as energy prices escalated. Figure 4–9 has yet another update to accommodate 21st-century prices. For an example of how to use this nomogram, find the prices of propane and heating oil equivalent with natural gas at $4.00 per million Btu ($4.00/MMBtu):

> Draw a straight line from the left apex to $4.00 on the natural gas scale. Reading off the propane and heating oil scales: $0.34/gallon and $0.51/gallon are equivalent to $4.00/SCF natural gas and $3.81/MMBtu.

Any of the fuels can be similarly be related as long as the straight line starts at the left-hand side of the peephole.

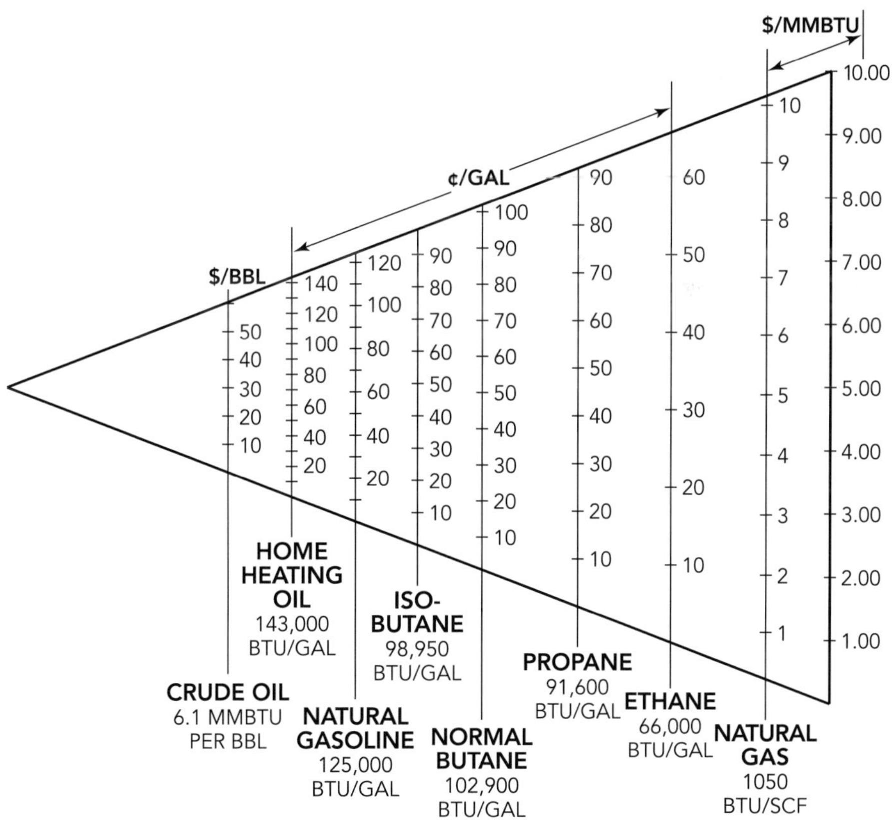

Fig. 4–9. Fuel value nomogram

How much?

Inevitably someone will want to know, "How much NGL is there in natural gas?" The answer is always, "It depends."

Every natural gas field has a unique composition of methane and NGLs, as well as some other, more or less undesirable compounds. Table 4–4 shows some approximations that cover many fields. *Dry gas fields* contain almost all methane and little NGLs. The so-called *liquids rich* production coming from shale gas fields is lower in methane and higher in NGLs. Natural gas fields in Pennsylvania, for example, have both kinds. In the north central and northeast part of the state, the bulk of the natural gas produced is dry. The natural gas production from the Marcellus shale is one of the natural gas streams most heavily laden with NGLs, particularly ethane.

Table 4–4. Natural gas composition mole content*

	Typical	Range
Hydrocarbons		
Methane (C_1)	93.0	80.0–98.0
Ethane (C_2)	3.4	0.5–10.0
Propane (C_3)	1.5	0.2–4.7
Isobutane (iC4)	0.4	0.05–2.0
Normal butane (nC_4)	0.5	0.06–2.0
Isopentane (iC_5)	0.2	0.03–1.00
Normal pentane (nC_5)	0.1	0.0–0.4
Hexane and heavier (C_6+)	0.3	0.0–0. 8
Inerts		
Nitrogen (N_2)	0.3	0.1–0.9
Helium (He)	Trace	0.0–2.0
Impurities		
Carbon dioxide (CO_2)	0.6	0.1–1.9
Hydrogen sulfide (H_2S)	Trace	0.0–0.2
Oxygen (O_2)	Trace	0.0–3.0
Water (H_2O)	Trace	0.0–0.01

*The engineers who calculate the NGL content in natural gas use mole percent instead of volume percent. The latter is more or less useless in calculating the outturn of NGLs from a gas plant because each NGL has a different conversion as it goes from gas to liquid, from the inlet to the outlet of the gas plant. The engineers have to use mole percents to get it right. Mole percents are rooted in the atomic weights of each of the NGLs. A mole of ethane has the same number of molecules in it as a mole of any other NGL, but has its own mole weight in pounds. None of this is intuitively obvious; see appendix 3 for a deeper explanation.

Commercial specifications

Propane. The Gas Processors Association (GPA), in concert with the American Society for Testing Materials (ASTM), has an established standard for the quality of propane sold to residential and commercial trade. The official name is *HD-5 propane*, an acronym for heavy duty—5% maximum propylene. Why the limit on propylene? About half the U.S. production of propane comes from refineries. Refineries make propane in their cracking processes. Along with the propane comes about 10 times as much propylene. Propylene is very reactive (see chapter 6, on refining, and chapter 8, on petrochemicals, for more details). In large enough concentrations, it can form gums and varnishes, which are bad news for users' equipment. So the HD-5 specification limits the propylene content to mildly benign concentrations to protect consumers. In contrast to residential and commercial use, the quality (presence of other stuff) of propane for petrochemical use is generally a negotiated item.

Other NGLs. Ethane, the butanes, and natural gasoline are used almost exclusively as feedstocks to make other products. The specifications for each are negotiated between buyer and seller, though common usage usually prevails. The same goes for propane used as petrochemical feedstock.

Sulfur out—sulfur in

Sulfur out. NGLs coming out of the ground typically have a few sulfur compounds such as hydrogen sulfide (H_2S) traveling along with them. Chapter 5, on gas plants, discusses how most of these compounds are physically separated from the NGL mix. Sulfur atoms embedded as part of the natural gasoline molecules are a bit more stubborn. The hydrocarbon molecules are larger and more complicated. Removing that sulfur from natural gasoline is usually relegated to a refinery. There, hydrotreaters use hydrogen and catalysts to chemically attack the sulfur-containing molecules. The hydrogen reacts with the sulfur, forming H_2S. In the process, the natural gasoline molecule is modified, usually breaking it where the sulfur atom was attached and making it a smaller molecule. Some natural gasoline can have as much as 2,000 parts per million (ppm) sulfur. The regulated specifications for motor gasoline in many places have a limit of less than 10 ppm sulfur. So the refiners have to work natural gasoline over to get it suitable for blending into motor gasoline.

Sulfur in. In the beginning, use of gaseous fuels like propane and natural gas in homes, farms, and commercial buildings presented a situation even more hazardous than today. Natural gas and propane are odorless. Leaks *have* to be easily noticed. After the 1937 natural gas explosion in a New London, Texas, school that killed 295 students and teachers (see fig. 2–4), governments mandated that natural gas and propane sold for retail fuel use at the burner tip or in an internal combustion engine had to have an odorant added. The most common additive is ethyl mercaptan (C_2H_5SH) a sulfur-containing compound. That is ironic since the propane will have been treated to remove sulfur (but of another form.)

The odorant is added sparingly because the human olfactory nerve has an extraordinarily acute sensitivity to various sulfur-containing odorants. It can detect a mercaptan concentration of only 3 ppm.

When the propane burns, releasing H_2O and CO_2, the small amount of malodorous mercaptan converts to H_2O, CO_2, and, SO_2 (sulfur dioxide). Sulfur dioxide itself smells like rotten eggs, but with just 1 ppm in the burned gases, this stinker is virtually undetectable.

Other odorants that are sometimes used include dimethyl sulfide, isopropyl mercaptan, secondary butyl mercaptan, tertiary butyl mercaptan, and thiophane. (Any dog owner whose pet has had an encounter with a skunk knows the smell of thiophane, which comes from the same mercaptan family.) Each of these compounds has slightly different properties, and blends are often used to accommodate longevity, freeze point, and other specifications.

Takeaways from the Chemistry and Physics of NGLs

- The NGLs are made up of paraffins, cyclics, and aromatics.
- Each NGL has distinctive chemical and physical properties that makes it suitable for different market applications.
- For safety's sake, retail propane has an odorant added so leaks can be more readily noticed.

Exercises

1. What are the six C_3's and C_4's?
2. Why are the natural gasoline properties of volatility, heating value, vapor pressure, and octane number always an approximation?
3. Why put sulfur compounds back in propane when the gas plant has just taken out sulfur compounds?

CHAPTER 5

Gas Plants

A process cannot be understood by stopping it.

Understanding must move with the flow of the process,

must join it and flow with it.

—Frank Herbert (1920–1986) *Dune* (First Law of Mentat)

Frac Spread

Anyone wanting to know whether gas plants are making money or not turns to calculations of the *frac spread.* That indicator, in general, shows the sales value of the NGLs produced at a gas plant less the value of the natural gas sold from that plant.

Why "in general"? Frac spreads have no official definition. Any frac spread is in the eye of the calculator. For example, some frac spreads take the difference between the natural gas price and recovered propane–butanes–natural gasoline prices. Others include ethane in the NGL part of the calculation.

Some published calculations use the price of propane as listed on the commodity exchanges as a surrogate for NGL prices because there are no totally transparent transaction prices for ethane, butanes and natural gasoline, only prices from firms that survey the market. And many frac spreads take measurements of NGLs at Mont Belvieu, Texas, and natural gas at Henry Hub in Erath, Louisiana. Geographic differences are sometimes dismayingly large.

The production engineers have pushed their rich gas—so named probably because of the money there—into a pipeline. In almost all cases, this gas has to be treated to reduce whatever contaminants are present. Even more interesting right now, most natural gas has enough natural gas liquids (NGLs) to warrant a processing facility to remove and recover them for sale.

Removal Imperatives

Typically, gas leaving the lease can have up to 15% NGL dissolved in it. Of course, natural gas streams are hardly ever typical. The NGL content can be as high as 25% or as low as 2% or 3%. Even for the typical stream, there are compelling reasons to separate the NGLs from the natural gas stream.

Value

Some or all of the NGLs—ethane, propane, butanes, and natural gasoline—are usually more valuable than natural gas in their own markets. Some NGLs could be left in the natural gas stream and sold at the natural gas price. Natural gas prices used to be based on volume, in dollars per thousand standard cubic feet ($/MSCF), but today are based on its energy content, that is, dollars per million British thermal units ($/MMBtu). NGLs have more Btu per cubic foot of gas than does natural gas. (See the price/Btu nomogram in fig. 4–9 in chap. 4.) The specifications for marketable natural gas (a maximum of about 1,100 Btu/SCF) do allow for some NGL content, but not a lot.

Rich gas, wet gas, dry gas

Some organizations use *wet* gas to mean gas with the NGLs still in it. Others use it to mean gas before the water vapor is removed and use *rich* gas for the NGL-laden gas. Conversely, *dry* gas could mean gas coming out of the gas plant with the NGLs having been removed, or it could mean dehydrated gas. Rich gas always means the NGLs are still in there.

If the NGLs can be removed at a reasonable cost, and usually they can, and the NGL markets are reasonably nearby, and they often are, the gas owners will build a gas plant or rent someone else's capacity in a tolling arrangement to recover the NGLs for sale.

Natural gas pipeline operations

Many natural gas production facilities are far removed from their ultimate markets. Take the example of a gas well offshore in the Gulf of Mexico. In the wintertime that gas moves by pipeline to landfall, and then as far as, perhaps, Belmont, Massachusetts. The route takes the gas along the Louisiana lowlands, up and across the Cumberland Plateau, over the Appalachians, through the Jersey lowlands, and over the Berkshire Mountains before it reaches its final destination. Along the way, the pipeline has scaled mountains and dropped into valleys.

In the cold winter months, gas temperatures often drop below freezing along the way. During these times, pipeline pressures can reach perhaps 2,000 psi in anticipation of having to increase deliveries during a cold snap. They can drop quickly to 1,000 psi as the draw from the pipeline feeds furnaces in homes and commercial buildings.

All that sets the stage for peril if the NGL content is too high. At higher pressures or colder temperatures, or both, the heavier NGLs dissolved in the natural gas can form droplets as the natural gas passes through its dew point. This dew point phenomenon is commonly observed on cold autumn mornings. As temperatures drop overnight, they reach a point when water vapor in the air (humidity) gets cold enough to condense, leaving glistening droplets on lawns and windshields. On even colder mornings, the water vapor becomes a hoary frost. Drivers have to scrap the ice from their windshields or turn on their car heaters; golfers have to wait around until the frost delay is lifted, lest they damage the frozen turf.

The NGL content of natural gas operates in the same way. The details of the relationship among dew point, temperature, pressure, and NGL content are complicated enough that they are relegated to appendix 1.

Passing through the dew point can result in several bad conditions. First, the accumulated liquids that have condensed take up space and cut down on the pipeline capacity. Second, when the pipeline operations call for a rapid increase in pressure, slugs of liquid NGLs can be pushed along the line. If those liquids enter the turbines in the next compressor station, the liquids can damage the turbine blades and cause the compressors to fail. Third, instruments, sensors, controllers, and safety devices can be clogged with liquid and rendered dangerously ineffective.

The natural gas pipeline companies try to deal with this problem by setting limits on the dew point of the natural gas they accept. That limits the amount of the heavier NGLs and even ethane and propane in those places where the NGL content is quite high, such as the natural gas coming from some shale gas formations.

The pipeline companies, in deference to consumers, also worry about the Wobbe index. When natural gas is burned, it has to be mixed with the right amount of air to get an efficient flame. Home appliances and industrial burners are set to expect a certain range of Btu contents of the incoming gas. The Wobbe index measures the match between the composition of a batch of gas and a standard burner tip design, which, of course, is fixed. The details of the Wobbe index are also complicated enough to deflect the discussion to appendix 1.

Contaminants

Natural gas sometimes contains unacceptable levels of the acid gases, carbon dioxide (CO_2), hydrogen sulfide (H_2S), and other sulfur compounds such as carbonyl sulfide (COS) and the mercaptans (RSH). If the water dew point is reached in a pipeline, the sulfur compounds can react in the condensed water with metals (pipelines, appliances, turbines) forming metal sulfides, that is, corrosion. Carbon dioxide also can react with condensed water, to form carbonic acid (H_2CO_3), which can likewise cause corrosion. Pipeline companies set maximum acid gas content to protect their facilities and their customers. The specification can vary, but generally calls for no more than 2%–4% of CO_2 and less than 0.25–1.0 grains per 100 cubic feet of total sulfur. (Grains? See the accompanying box.)

Barley, sulfur, and bullets

The grain is one of the oldest measures. Originally, as far back as the Bronze Age, the grain equaled the weight of a single seed of barley or wheat. More recently, Great Britain set the grain equal to 1/7000 of a pound in the Imperial Weights and Measures act of 1855, and reaffirmed it, together with the United States, in 1959. Only a few commercial operations use grains as a unit of weight, including the sulfur content of natural gas, rifle bullets, and precious metals used in jewelry.

Treating. Every gas plant has facilities for removing the acid gases, even the ones that do not remove any NGLs (fig. 5–1). *Conditioning*, and sometimes treating, refers to the removal of acid gases as well as the moisture (water) from the gas. *Sweetening* refers to removing only the sulfur compounds (desulfurization): hydrogen sulfide, carbon disulfide, carbonyl sulfide, and the mercaptans. *Dehydration* refers to the water-removal step. Many plants also have a nitrogen content reduction step.

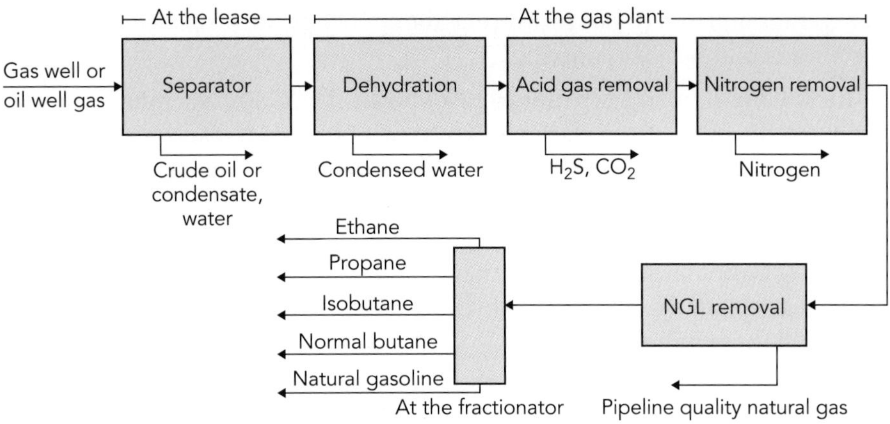

Fig. 5–1. Treating in a gas plant

Where there are small volumes of gas, the sulfur compounds can be removed by passing the gas through activated charcoal. For larger volumes, the most common process for treating natural gas uses amine treatment. The process is elaborate enough and important enough, but off the subject of NGLs enough, that it has been relegated to appendix 2.

NGL removal retrospective

Sure, natural gas plants condition gas, but their more interesting function (here) is separating the NGLs from the incoming gas. In the early days of the oil patch, most producers flared natural gas. The development of seamless pipe in the 1880s and reliably welded connections in the 1920s initiated interest in long-distance natural gas transmission lines. Along with that came the realization that the NGLs can condense in transit.

To remedy this, some pipeline companies forced producers to build gas plants and reduce the dew point of the raw natural gas, which was then about 50–80°F coming off the lease. (Recall that the hydrocarbon dew point is defined as the temperature at which the heavier components start to drop out of the vapor as liquids.) The initial technology of choice was compression: Increase the pressure to 100 psi and have the heaviest of the NGLs liquefied; then draw them off. That reduced the dew points to about 10°F and removed about 25% of the natural gasoline.

During the winter months, in-transit condensation still posed a problem, but the introduction of refrigeration to gas plant processing schemes dropped the dew point down in many cases to 0 to 2°F. Not only did that please the pipeline companies, but producers found themselves with marketable NGLs.

Refrigeration plants began recovering about 50% of the propane and most of the butane and heavier content.

Increasing demand for motor gasoline in the 1930s led refiners to call for more normal butane. During World War II, the demand for isobutane spiked as the military called on refiners to make high-octane aviation gasoline components by reacting isobutane with butylene and propylene. In response, newly improved refrigeration designs were enabled by the ironically cannibalistic use of ethane and propane as refrigerants. That allowed gas plant operations to achieve even lower temperatures and recover almost all the butanes.

In the 1960s, offshore production of natural gas in the Gulf of Mexico began. The rich NGL-laden gas came ashore by pipeline near the petrochemical industry locations in Louisiana and East Texas. Large *straddle plants* in Louisiana, so-called because they sat on those pipelines as they came ashore, processed this gas. Gas plant builders chose yet another new technology, lean oil-absorption, to make deeper cuts into the ethane fraction to satisfy the petrochemical industry's appetite.

In 1973, the Arab oil embargo caused oil prices to spike. (In the United States oil prices increased from $3/barrel to $12.) But natural gas prices, still under U.S. federal regulatory control, stayed constant at only about $.25/MSCF. That gave gas producers a huge incentive to switch new gas plant construction to the turboexpander technology to capture 60% to 90% of the ethane and virtually all the propane, butanes, and natural gasoline. The growing petrochemical industry continued to provide a ready home for the ethane.

Starting in the 1990s, decontrolled natural gas prices sometimes spiked and cratered, leaving investors uncertain about the outlook for ethane recovery economics. Gas producers still had to meet the dew point specifications, particularly at locations remote from petrochemical sites. They turned to less expensive but newly improved refrigeration designs. Other special situations, mostly small volumes of natural gas, used the less costly Joule-Thompson cycle plants, a simplified version of a refrigeration plant. Some were mounted on skids for easy, quick installation.

Current Configuration

At last count in the United States, there were nearly 200 turboexpander plants, more than 80 refrigeration plants, still more than 60 older absorption plants, and a handful (about 10) of Joule-Thompson plants.

Refrigeration plants

These facilities (see fig. 5–2) take advantage of the propane and sometimes ethane as refrigerants. Just like the Freon in an old home air conditioner, ethane and propane can be compressed, causing it to condense to a liquid state. It is then *autorefrigerated* by release of the pressure, often termed throttling. This "flash" autorefrigeration step drops the temperature precipitously to –10°F to –35°F as some of the liquid vaporizes. The resulting cold gas/liquid mixture is separated. The cold liquid refrigerant stream cools the incoming NGL-laden natural gas stream, causing much of the propane and most of the heavier NGLs to liquefy and drop out. The liquid refrigerant vaporizes in the process and then joins the other refrigerant vapor stream to be compressed and liquefied to start the cycle again.

Some newer refrigeration plant designs can recover some of the ethane and virtually all the propane and heavier NGLs.

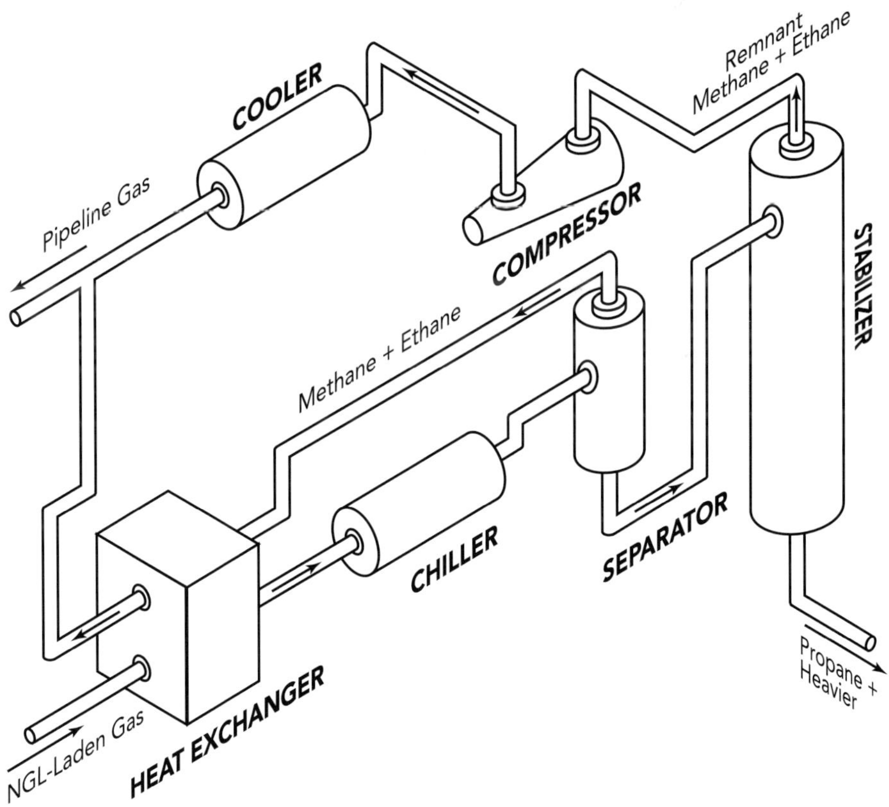

Fig. 5–2. Refrigeration gas plant

Lean oil-absorption plants

Methane is so volatile that separation by conventional distillation (described later) would require extensive facilities and energy to split out a high percentage of NGLs. An early option allowed that. The curious but key piece of equipment in this facility is the *absorber*, an unusually shaped column with perforated trays or loose packing in it (fig. 5–3). It uses a solvent called *lean oil* to absorb the NGLs out of the natural gas.

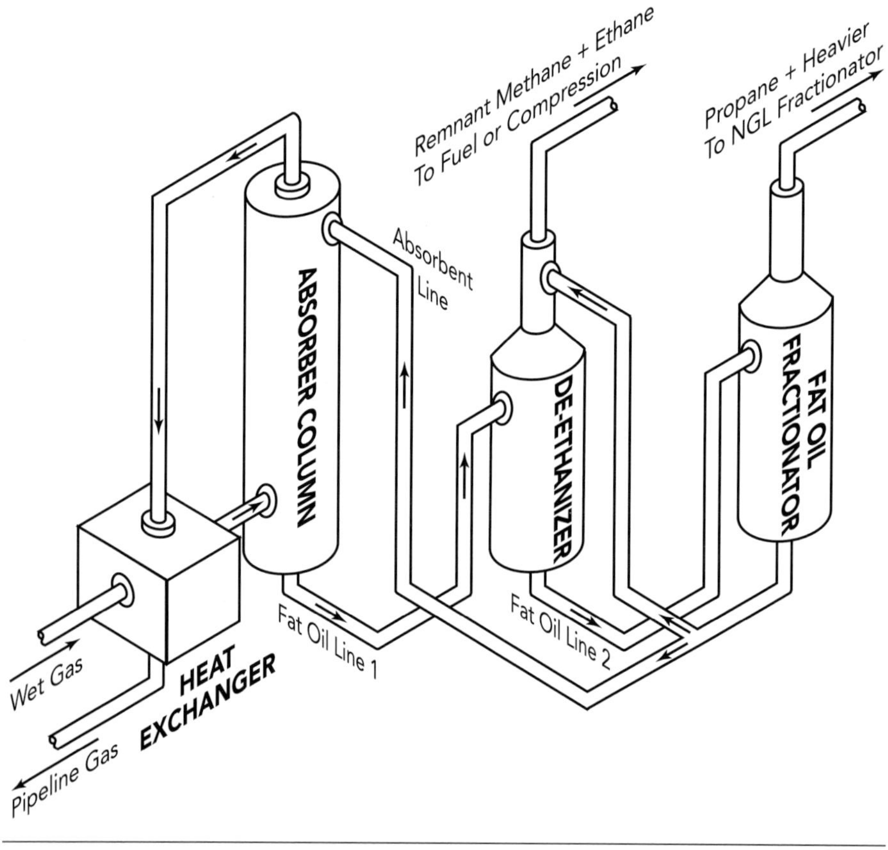

Fig. 5–3. Lean oil-absorption plant

The NGL-laden natural gas enters the bottom of an absorber as a gas and rises due to gravity. The lean oil, a hydrocarbon anywhere in the range of C_7H_{16} to $C_{12}H_{26}$, enters near the top of the column as a liquid and flows downward. As the NGL-laden natural gas and lean oil randomly slosh past each other, the lean oil absorbs the natural gasoline, the butanes, and some propane. Some small amount of ethane may come along, depending on the absorbing solvent and the operating conditions. But the methane is too volatile to be absorbed. It continues up and out the top of the absorber as a vapor with just a small amount of clingy NGLs still dissolved in it.

From the bottom of the column comes *fat oil*, a mixture of the lean oil and the absorbed NGLs. The mixture is heated or the pressure is reduced as it enters a fractionator. The solvent has a high enough boiling temperature that it is easily separated from the NGLs in a fractionating column and comes out the bottom as reconstituted lean oil. The lean oil stream goes back to the absorber to do continuous recycle duty. The stream from the top of the fractionator, the NGLs, eventually goes to fractionators to be separated, one from other (more on this later).

Turboexpander plants

Most new plants of any size use technology introduced in 1964, with improvements in mechanical design ever since. The heart of the plant is the turboexpander, a device that dramatically super-expands the incoming NGL-laden natural gas. As it expands, the temperature drops, sometimes as low as –200°F. That makes the words *cryogenic plant* interchangeable with *turboexpander plant*. At those low temperatures, as much as 98% of the NGLs, even the ethane, condense and separate from the methane.

In the simplified diagram of a turboexpander plant (fig. 5–4), the flow starts with cooling the NGL-laden natural gas to about –60°F in a heat exchanger. At that temperature, most of the NGLs liquefy. In a separator, just a big tank, the liquid NGLs drop to the bottom and are taken off. (The liquid still has a little methane dissolved in it, so it goes to the fractionator.)

The natural gas vapor exiting the top of the separator has some NGLs dissolved in it, mostly ethane, and some propane and butanes. That gas stream goes through the turboexpander. The gas expands and the temperatures goes from low (–60°F) to very low (–90°F to –200°F). That causes almost all the NGLs—but no methane—to liquefy. A fractionating column, the *demethanizer*, allows orderly separation of the gaseous methane from the liquefied NGLs from the turboexpander. The NGLs from the separator (above) entering the lower part of the column have just a little methane left in them, so the demethanizer deals with that, also. The NGLs exit the bottom of the column and are heated to ambient temperature.

The methane coming from the top of the column has to be warmed in a heat exchanger and then compressed to meet natural gas pipeline specifications. The compressor is in this case is mechanically connected to the turboexpander by a common shaft. The power from the expanding gas in the turboexpander drives that compressor. Another motor-driven compressor is usually needed to get the methane up to pipeline pressure, about 1,000 psi.

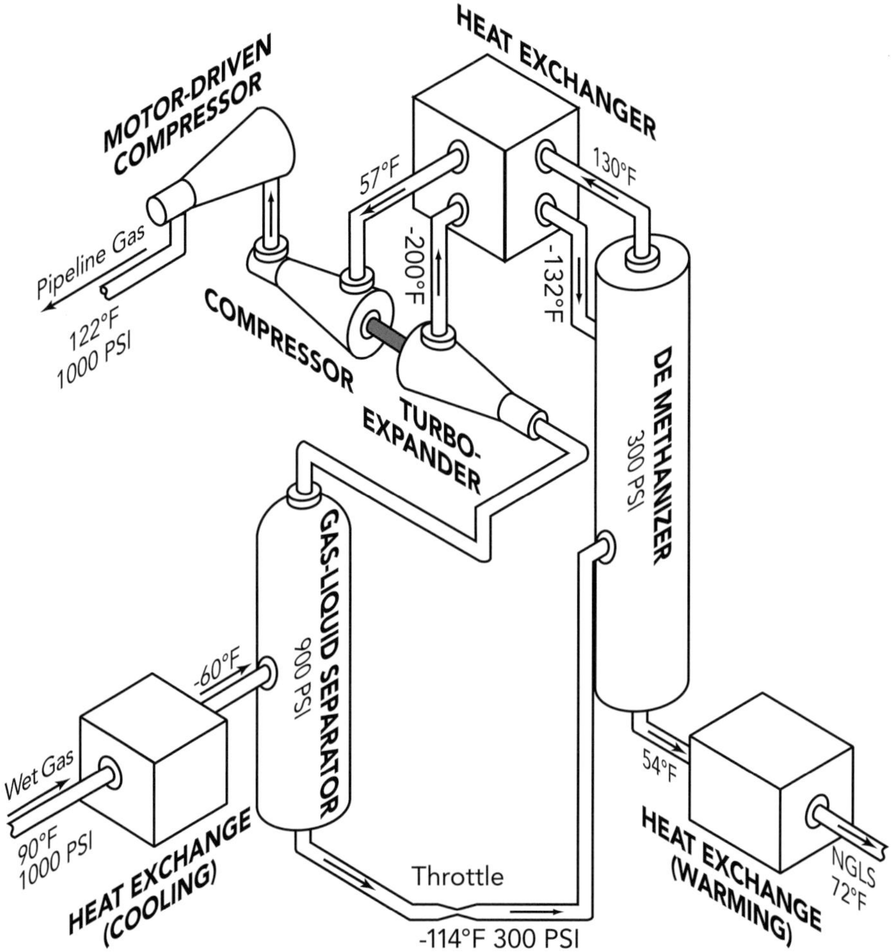

Fig. 5–4. Turboexpander plant

The NGL-laden natural gas may come into the plant warm because it has previously been compressed; relatively warm natural gas leaves the plant. In between, the heat from the incoming gas warms the outgoing gas; the cold stream from the turboexpander cools the incoming gas further (not shown.) Figure 5–4 shows two different heat exchangers, but in a well-integrated turboexpander plant, all this frantic heat switching takes place in one piece of equipment called the *cold box.*

Fractionation

After NGL separation by a refrigeration plant, a turboexpander, or a Joule-Thompson plant, the stream of mixed NGLs needs to be separated into its various parts. Sometimes a fractionation column sits right at the gas plant. In other situations, the mixed NGLs are moved to another site, sometimes hundreds of miles away, by pipeline as a *raw make* or *Y-grade* stream for separation. In either case, the workhorse in this process is the fractionator, also known as a distilling column or a splitter.

The fundamentals of fractionation (or distilling) are evident from any moonshiner's operation (fig. 5–5). Into the fermenting vessel go corn mash, some water, and some enzymes. After a few days, the starch in the corn mash chemically morphs into ethyl alcohol. When that reaction has sufficiently completed, heating the liquid mixture accomplishes the separation. The boiling temperature of alcohol (about 173°F, depending on how much water comes with it) is lower than the boiling temperature of water (212°F). So heating the contents to about 200°F causes the alcohol to boil—go from liquid to vapor. A little water (about 5%) always comes along with it. The vapor rises and goes through a through a cooling stage, where it condenses back to liquid, that is, almost neat ethyl alcohol.

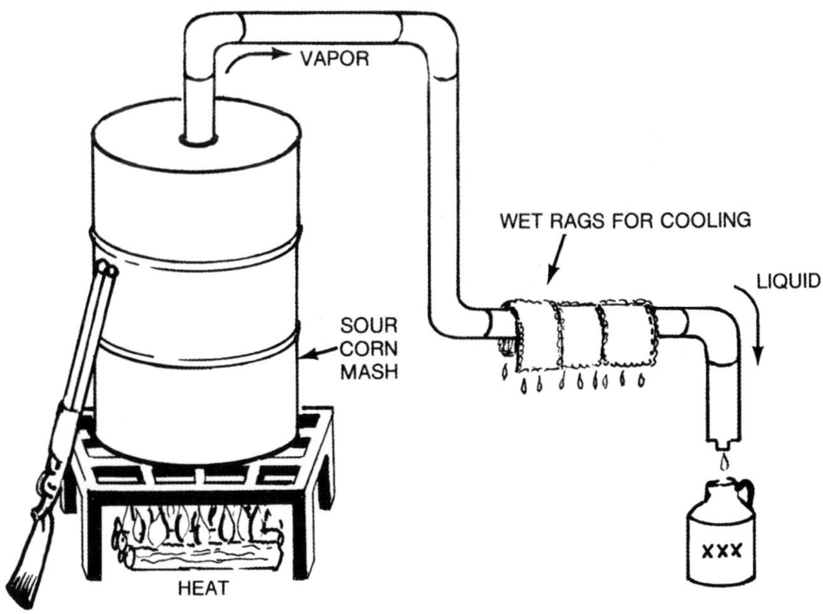

Fig. 5–5. Making moonshine

The essential four steps of distillation are the following:

- Heating
- Vaporizing
- Cooling
- Condensing

Modern distilling operations in the refining and gas plant industries have not used use this *batch method* since the 1920s. Instead they use distillation columns that have in them either multiple perforated trays or packing (fig. 5–6). Packing can be in the form of loosely loaded, small, specially shaped, metal pieces sitting on perforated trays or even corrugated sheets in vertical layers.

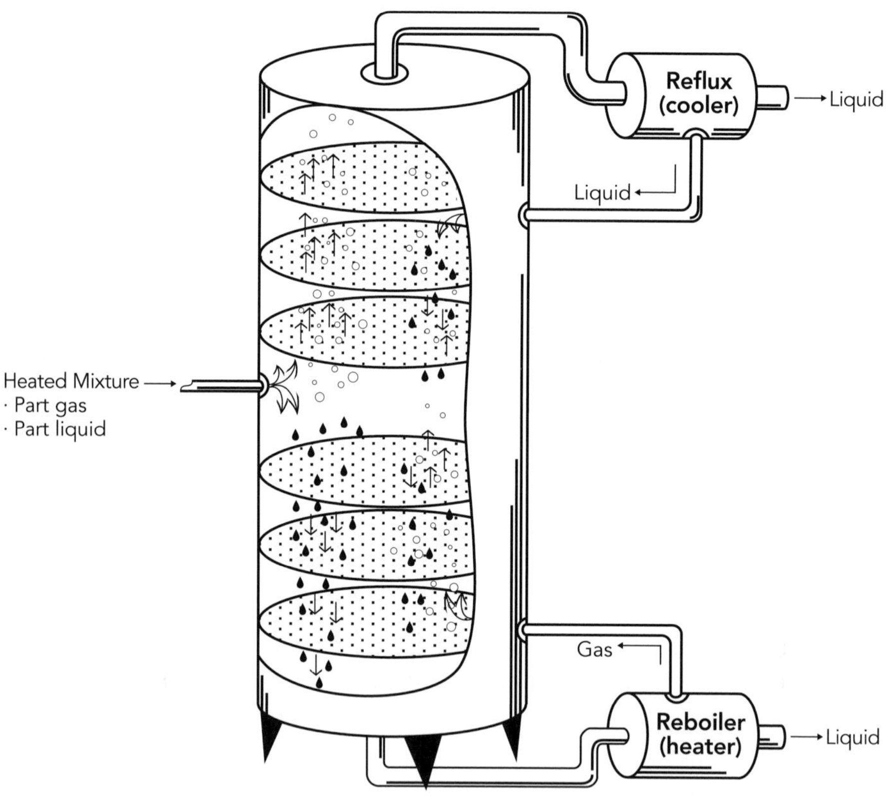

Fig. 5–6. Fractionator with trays, reflux, and reboiler

To run a fractionating column, the incoming feed must be heated so that it is part liquid and part vapor. NGLs usually arrive as a liquid under pressure. Thus lowering the pressure or heating them on the way into the column will permit lighter components to vaporize. As the gas/liquid mixture enters the column, gravity moves the vapor up the column and the liquid down.

As the vapor moves up the column, its temperature drops, causing some of it to condense on the trays or packing. The trays overflows or the liquids drip from the packing. Either way, the condensed liquid moves down toward the bottom. Meanwhile, by the time the vapor reaches the top of the column, almost all the heavier (higher boiling temperature) NGLs have liquefied, dropped out, and headed downward. The gas coming out the top of the column is mostly ethane.

Some propane and butane might make it out the top with the ethane, so the stream goes through a *reflux* stage, a cooler that condenses the stream back to liquid. Some of that liquid, sometimes as much as 90%, is reintroduced back into the column near the top and starts working its way down, again by gravity. It meets hot vapor coming up; almost all the reflux stream turns back into vapor and goes out the top. But the heaviest part, that is, any propane and butanes, works its way down the column as a liquid. Using this reflux technique, the ethane stream achieves a high purity because it has been redistilled many times.

Out the bottom of the column come the propane, butanes, and natural gasoline as a liquid. Since it may have a little ethane still dissolved in it, the stream goes through a heater (*reboiler*) that vaporizes some of it. The vapor reenters the bottom of the column. It meets liquid propane, butane, and natural gasoline coming down. As the liquid and vapor pass by each other, the ethane continues up the column, but the other three condense again to liquid. That scrubbing eliminates all but a trace amount of ethane from the bottom stream.

Now the ethane has been removed, and there is a stream of mixed NGLs, propane, butanes, and natural gasoline. Another three columns (fig. 5–7), four in total, are needed to separate the five streams completely. The stream from the first column, the deethanizer, goes to another fractionator, a depropanizer, and then a debutanizer. The butanes can be separated from each other in a deisobutanizer. In each case, the stream has to be heated somewhat (or the pressure dropped) to ensure that the liquid bottoms from one column go into the next as a vapor/liquid mix. The sequence necessarily follows the boiling points of each of the NGLs.

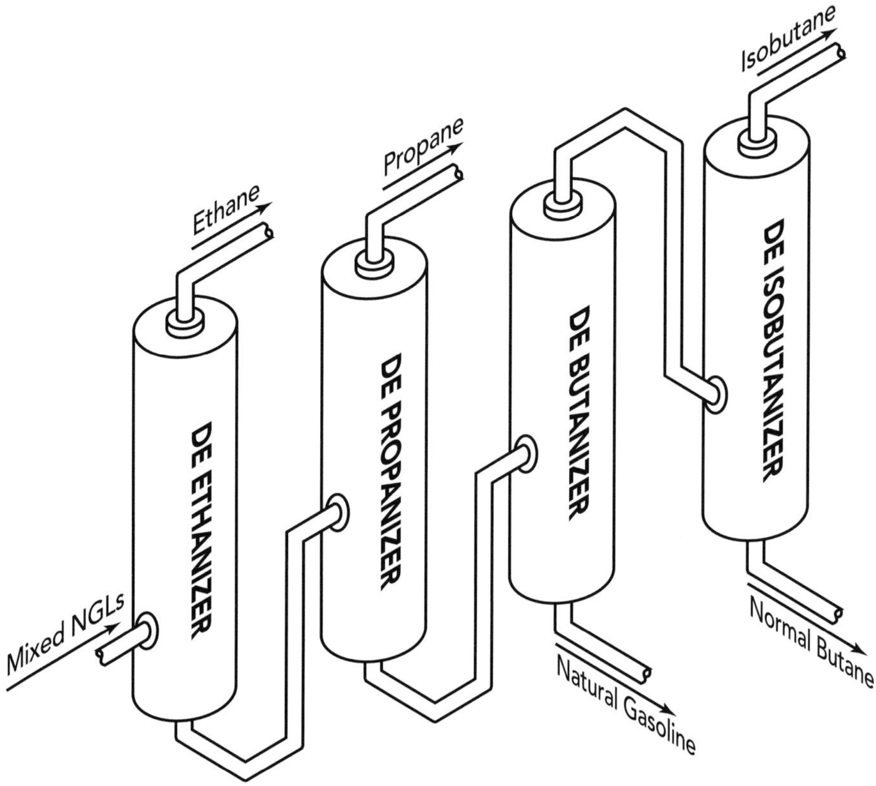

Fig. 5–7. NGL fractionation

This description of distillation might seem complicated. Even so, it has been inordinately simplified and shortened here (so much so it might make a chemical process engineer guffaw.) Still, the principal takeaways from any discussion of distilling are:

- Vapor/liquid goes into the column.
- Vapor comes out of the top.
- Liquid comes out of the bottom.
- Heating, vaporizing, cooling, and condensing is what fractionation is all about.

Storage

The fractionated gaseous ethane may be directly pipelined to the end user or stored in pressurized or refrigerated tankage for eventual transport. The fractionated propane and normal and isobutane products typically undergo further desulfurization and dehydration steps to meet product specifications.

(See appendix 2.) Natural gasoline goes to tankage not requiring pressure or refrigeration. It still has to be a closed system to prevent vapors escaping to the environment. Much more on this is covered in chapter 7.

Takeaways from Gas Plants

- Gas plants remove the NGLs from the natural gas streams and condition or treat the resulting products to lower contaminant (typically sulfur compounds, carbon dioxide, and water) levels to meet industry specifications.
- Contemporary gas plant designs include refrigeration plants (old and new technology), lean oil absorption plants (old technology), and cryogenic turboexpander plants. In that order, they recover increasingly higher percentages of the NGLs in the natural gas.
- NGL separation from each other can take place at the gas plant or at a remote fractionator. To separate the NGLs, they go through heating, vaporizing, cooling, and condensing steps repeatedly in several columns.
- The treating steps typically remove sulfur compounds (desulfurization), carbon dioxide, and water (dehydration).

Exercises

1. Name the economic and operational rationales for having gas plants.
2. What are is the latest technology used in gas plants, and what did it replace?
3. What are the differences among conditioning, sweetening, and treating?

CHAPTER 6

Refineries and the Unnatural Gas Liquids

Only God can make a tree, probably because it's so hard to get the bark on.

—Woody Allen, *Without Feathers*

Fire at the SMR!

At Shell's Wood River Refinery on July 2, 1965, the klaxons sounded, and firefighters and staff hurried toward the steam methane reformer (SMR). Foreman Kenny Baird and 26-year-old Bradley Peterson arrived at the SMR to find a hissing sound, frost forming on a valve, and a pile of pallets burning about 15 feet away. Peterson moved to turn off the valve.

Baird grabbed him and told him to get a broom and hurry. Puzzled, Peterson left and returned in a minute with the broom. Baird held the business part of the broom out in front of him and walked toward the valve. About 10 feet away, the broom burst into flame.

"Run to the control room and tell them to vent the A83 line to the flare," he shouted to one of the arriving firefighters, while Peterson stood there astonished.

"The escaping hydrogen made that noise and dropped the temperature of the escaping hydrogen around the valve to make that frost," he said to Peterson. "Only the front of the hydrogen cloud was burning, giving off scorching heat. But when hydrogen burns, it has no visible flame. That's why it seemed that only the trash over there was burning and nothing else. I just saved your life, Peterson."

The Unnatural Gas Liquids

Just as there are regular tomatoes and organically grown tomatoes, there are natural gas liquids and unnatural gas liquids. Almost all the ethane, propane, and butanes that come about in refineries are not natural. They are created in the process of turning crude oil into refined products—gasoline, jet fuel, diesel and furnace oil, heavy fuel oil, lubricants, and petrochemical feedstocks (fig. 6–1). Refining is a complex affair, and the only way to find out where the (unnatural) gas liquids come from is to go through the process step-by-step.

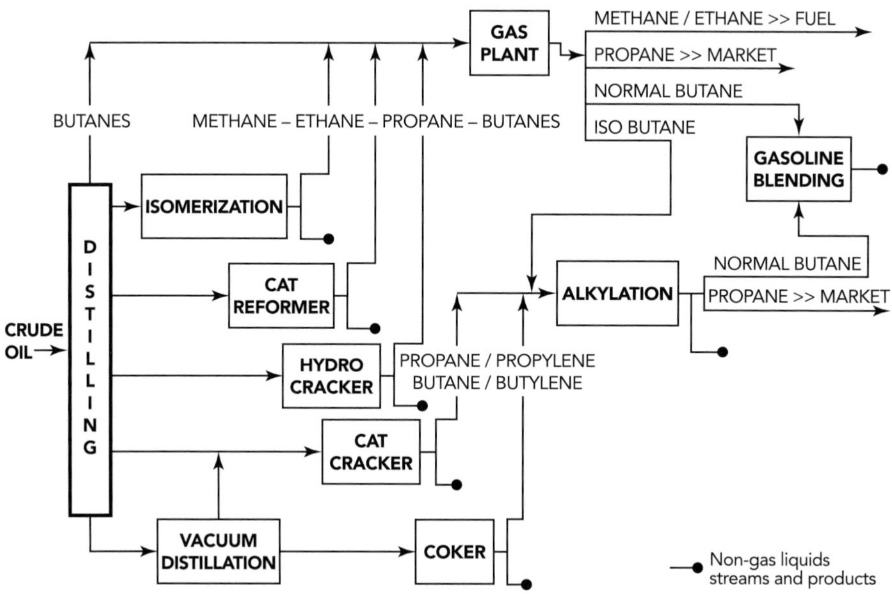

Fig. 6–1. Refinery sources and uses of NGLs

This chapter focuses mostly on propane and butanes. Ethane is created in several process units, but in most refineries, it is not recovered for sale. Ethane stays in the stream along with methane, which goes ultimately to the refinery fuel system that burns it. Refineries have huge appetites for fuel to make steam or run process unit furnaces and heat exchangers.

The words *natural gasoline* do not appear in this chapter until the subject of treating and gasoline blending come along. The refinery equivalent of natural gasoline is straight run gasoline. It comes directly out of the crude oil distillation unit, hence the term *straight run*. It is more or less interchangeable with natural gasoline—same chemical constituents and properties, same processing schemes.

The units in the refinery that involve gas liquids fall into five categories:

- Separation
- Crackers
- Rearrangers
- Treaters
- Blenders

Only the first four make gas liquids. The fifth uses butane and natural gasoline.

Separation

Crude oil distillation unit

The crude oil distillation unit in a refinery, shown on the left of figure 6–1, separates the incoming crude oil into various fractions or cuts. Each one has hydrocarbons with molecules that boil between specific boiling temperatures. That is how distilling units work—by dealing with boiling temperatures in the same way as was explained in chapter 5, on gas plants. All distilling units or fractionators take in a mixture of vapor and liquids. The liquids fall toward the bottom of the column, and the vapors work their way to the top. The trays or packing in the column act as multiple distillation steps as the vapors move up and the liquids move down, passing and scrubbing each other over and over to ensure that the correct molecules go all the way up or all the way down.

The crude oil distilling column in figure 6–1 has not only vapors coming out the top and liquids out the bottom, but also *side draws*. From those outlets, intermediate boiling-temperature fractions emerge. How does that happen? After all, didn't the description that went along with figure 5–5 have only tops and bottoms, nothing coming out in between? It is easier to think of a crude oil distillation column with side draws as a set of five or six distillation columns in series, with tops and bottoms coming out. But in this case, they are stacked on top of each other, so the tops of one go directly into the bottoms of the one above it, and the bottoms of one go . . . well, you get it.

The role

The crude distillation unit may be called the refinery postmaster. It sorts the molecules and sends them to be dealt with in different processing units downstream in the refinery. The typical outturn from the distilling unit would go as follows:

- Light gases (butanes and maybe a little propane) to the refinery gas plant
- Straight run gasoline to treating and then to gasoline blending
- Straight run naphtha to catalytic reforming for upgrading to gasoline
- Kerosene to hydrotreating to jet fuel
- Straight run light gas oil to treating to diesel and home heating oil (known as *distillate fuels*)
- Straight run heavy gas oil to catalytic cracking or hydrocracking
- Straight run residue to vacuum flashing

The only significant natural gas liquid (NGL) that comes from the distilling unit is butane. Often oil producers or pipelines put extra butane in crude oil as it moves in a pipeline. The term *diluted crude* indicates that butane has been added up to the vapor pressure limit of the pipeline. When that diluted crude oil arrives at the refinery, the butane comes out the top of the crude oil distillation column and goes to the refinery gas plant (described later in this chapter).

Cracking Units

Any sophisticated refinery has a *catalytic cracker* or a *hydrocracker,* or both (fig. 6–2). Many also have cokers. Each of these units takes a stream of heavy oil and breaks up the complex molecules in it by actually cracking some of the bonds that hold some of the atoms together. Catalytic (cat) crackers and hydrocrackers take a stream (gas oil) from the distilling unit or *flasher tops* from the vacuum distillation unit (sometimes called the *vacuum flasher.*) The purpose is to turn those heavy, lower valued streams into gasoline and distillates, which are lighter, higher valued steams. Cokers take in the heaviest streams, the flasher bottoms. They crack them into the higher valued products as well.

During the cracking process, most of the heavy molecules break into those preferred gasoline and distillate molecules. A small but not insignificant portion of the cracking process results in molecules in the NGL range—ethane, propane, and butanes (C_2, C_3, and C_4's).

Because of the way they run, the composition of the outturn from cat crackers, hydrocrackers, and cokers all differ.

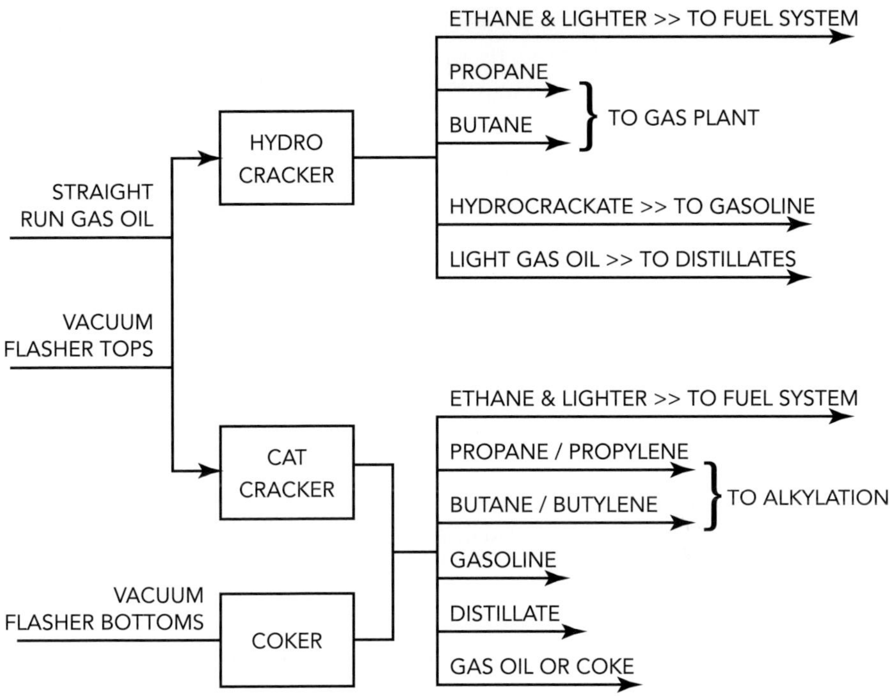

Fig. 6–2. Cracking units

Cat crackers

Here is the case alluded to in chapter 4, on chemistry, where there are not enough hydrogen atoms around to satisfy the valence of carbon. When molecules crack, there is a shortage of hydrogen to fill the places in the molecule where the cracking took place. Some of the newly created molecules go wanting, resulting in molecules such as ethylene, propylene, and butylenes, those reactive molecules with the double bonds. Ethane, propane, and butanes are created as well, as do larger molecules with double bonds.

In most refineries, the amount of ethane and ethylene from a cat cracker is too small to make it worthwhile to separate it from the methane that is also produced in the crackers. The three compounds, methane, ethane, and ethylene, are typically burned together in furnaces as refinery fuel—raising steam or generating heat.

The propane, propylene, butanes, and butylenes are separated from each other and usually put to good use:

- Propane is separated and sold to residential and commercial markets or to petrochemical companies as feedstock.

- Propylene can be used in a refinery alkylation plant where it is reacted with isobutane to make a high-octane gasoline-blending component; or it is sold to petrochemical companies as an important base chemical.
- Normal butane is separated and used as a blending component in gasoline or sold to petrochemical companies as feedstock.
- Isobutane can be used in a refinery alkylation plant where it is reacted with propylene and butylenes to make a high-octane gasoline-blending component called alkylate. In a few refineries, isobutane is not separated from normal butane and is used where normal butane is.
- Normal and isobutylene go to the alkylation unit, just as propylene does. There they react with isobutane to make isoheptane and isooctane, two high-octane gasoline blending components. Lesser amounts of isobutylene and normal butylene go to the petrochemicals industry as raw material.

The propane/propylene stream coming from a cat cracker is sometimes sold directly to the petrochemicals industry. Indeed, the petrochemicals industry gets almost half its propylene from refineries. The rest the petrochemical producers make themselves.

The propane/propylene stream from refineries is called *refinery grade propylene* and has about 40% propane along with it. Only a few petrochemical companies use refinery grade propylene in their processes. Most petrochemical process units need a higher concentration propylene called *chemical grade propylene* that has about 5%–8% propane in it. Other petrochemical processes use *polymer grade propylene,* which has less than 1% propane.

Petrochemical companies and some independent processors put refinery grade propylene in a fractionator to make chemical grade or polymer grade by removing propane. They do the same sometimes to upgrade the propylene that comes directly from olefin plants that crack propane, butane, or naphtha. The propane that gets removed in these fractionation operations goes into the general propane markets as purity propane or petrochemical feedstock.

Hydrocrackers

These crackers operate a bit differently from cat crackers—different catalysts, much higher temperatures and pressures, and, importantly, they add voluminous amounts of hydrogen to the reactors as the cracking takes place. With all that hydrogen around, as the molecules crack, the newly created molecules have no difficulty finding hydrogen partners. So the by-product C_3 and C_4's and the main products all have single bonds only, no double bonds. No ethylene, propylene, or butylenes come from hydrocrackers. Disposition of the C_3 and C_4's is the same as before—propane markets, gasoline blending, or petrochemical feedstocks.

As the hydrogen is added, and as the larger molecules crack, smaller but less dense molecules are created. As much as 15% more volume of products comes out of the hydrocracker than the volume of the feed that goes in—same weight, but more volume. Smaller molecules have fewer pounds per gallon than larger ones. Some refiners chuckle and call that "fluffing up the barrel."

Cokers

Many refineries have cokers that crack the "bottom of the barrel," the residue from the bottom of the crude distillation unit or the flasher bottoms from the vacuum flasher. Cokers operate without catalysts. They just heat up the very heavy feedstocks to about 1,200°F, put them in a large vessel called a coke drum, and let them cook. The large, complex molecules crack into smaller, simpler ones in the gasoline and distillate range. Cokers also make a lot of coke, a coal-like substance of nearly pure carbon. But like the cat crackers, there are lighter by-products—propane/propylene and butane/butylenes—and they are handled the same way as the by-products from cat cracking.

The Rearrangers

Some units in refineries do strange chemistry to some molecules. Because they shake up the molecules rather profoundly, they all have some by-product ethane, propane, and butanes. As before, ethane and lighter usually go to the fuel system.

Catalytic reforming

This process works on naphtha, a fraction somewhat heavier than gasoline. Naphtha, like natural gasoline, generally has a low octane number. That makes it unsuitable for blending directly into gasoline. A cat reformer converts this low-octane gasoline material to a high-octane gasoline blending component. Catalytic reforming does this by, well, reforming the "octane-challenged" molecules into higher octane ones. Straight chain paraffins convert to branch chain paraffins or to naphthenes, the cyclic compounds (fig. 6–3); naphthenes convert to aromatic compounds such as benzene and toluene. Each of these changes raises the octane number of the naphtha.

Typically, the octane number of the naphtha goes up 10 to 15 numbers. And the name changes from naphtha to *reformate.*

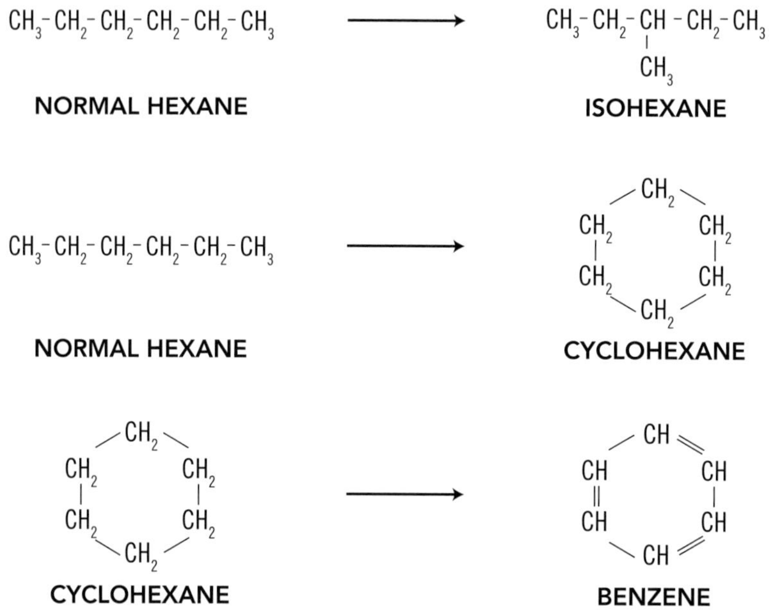

Fig. 6–3. Cat reformer molecular changes

All this transformation comes from the near magic of a catalyst, a platinum-palladium combination. Meanwhile, because of all this roughing up, by-products, methane, ethane, propane, and butanes (call them butane and lighter from now on) get created. They account for about 6%–10% of the outturn. No double-bonded olefins are created because the cat-reforming reactors have excess hydrogen being made in the process, especially as paraffins convert to cyclic compounds. A C_6H_{14} paraffin (hexane) might go to a C_5H_{12} cyclic (cyclohexane), and then to C_6H_6 (benzene) with three H_2 molecules left over.

Isomerization

As it turns out, isohexane and isoheptane have much higher octane numbers than normal hexane and normal heptane. An isomerization unit uses heat, pressure, and a platinum-palladium catalyst similar to the cat reformer's to make the straight chain paraffin into a branched paraffin (fig. 6–4). There are no perfect processes in a refinery, so a small percentage of by-product ethane, propane, and butanes are generated in the process.

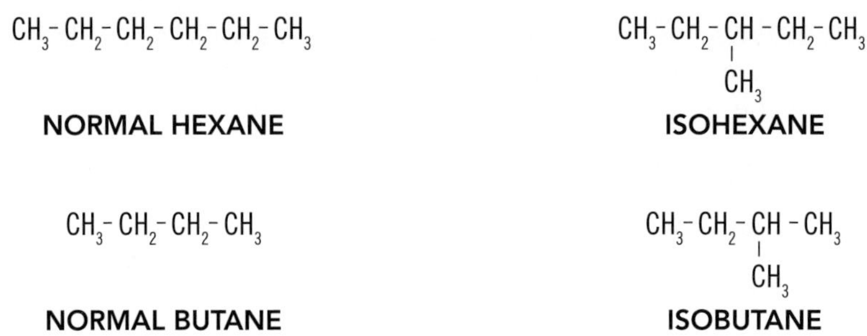

Fig. 6–4. Isomerization

Some refineries also have *butane isomerization* units that turn normal butane into isobutane. These refineries may be geographically challenged, unable to get a reliable, supplemental supply of isobutane for their alkylation plant operation. In that case, if they have enough normal butane around, they may want to have a captive supply of isobutane by building a butane isomerization (BI) plant.

Alkylation

This unit might be called an "uncracker." It takes the cracked by-products, propylene and butylenes, from a cat cracker or coker and reacts them with isobutane molecules. The molecules go from C_3, and C_4's to C_7's and C_8's, called alkylate (fig. 6–5). That way, the very volatile propylene and butylenes become larger, less volatile molecules and can be used in gasoline. They also happen to have very high octane numbers. The catalyst in the alky plant is either sulfuric acid or hydrofluoric acid.

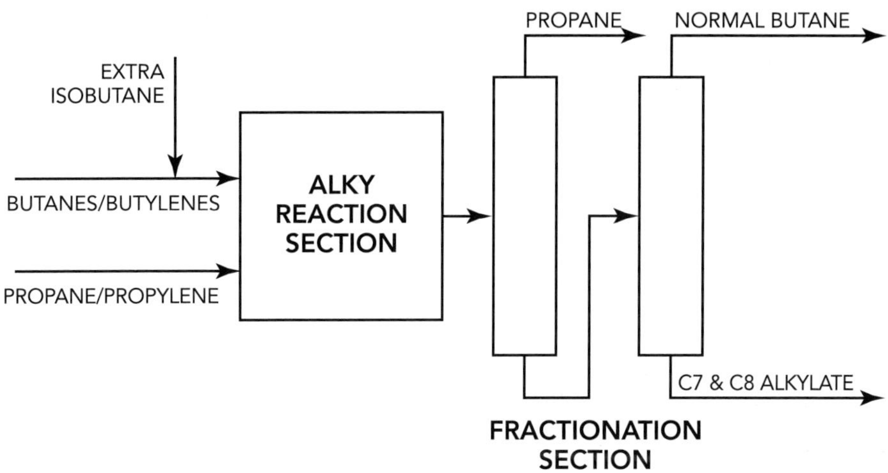

Fig. 6–5. Alkylation plant

The propylene and butylene feed (those reactive olefins with the double bonds) to the alkylation plant from the cat cracker and coker typically carry along the propane and the butanes. Those paraffins do not react in the alkylation process and are more easily separated after going through the process than before. Again, going into the alky plant are propane/propylene and butane/butylene; coming out are propane, butane, and C_7 and C_8 alkylate. The C_7's and C_8's have much higher boiling temperatures than the C_3 and C_4's (210°F and 211°F for the C_7's and C_8's versus 32°F or lower for the C_4's and C_3). That makes for an easier split in a fractionating column. Thus the alkylation unit is also a kind of gas plant that yields alkylate and both propane and normal butane. (The isobutane is consumed in the unit.)

Refinery Gas Plants

The by-product butane and lighter gas streams from the crude unit, cat reforming, and the hydrocracker go to the refinery gas plant to get separated from each other. The by-product propane/propylene and butane/butylene streams from the cat cracker and coker do not go there. They are handled in the alkylation plant, unless, as in a few refineries, there is no cat cracker or coker or alkylation plant.

Some refinery gas plants used same turboexpander technology as in natural gas processing plants. However, the percent of the lower boiling-temperature methane and hydrogen in refinery by-product gas streams is much lower, perhaps 5%–15%, compared to natural gas (85%–95%.)

Many refinery gas plants use absorption technology to separate propane and butane from the methane and hydrogen (fig. 6–6). In the first step, a compressor increases the stream pressure to about 200 psi. That causes almost all the propane and butanes to liquefy. Despite that pressure, some ethane and methane might still be dissolved in the liquid, just as a beer left on a bar for two hours still has just a little carbonation left in it. Later, after separating the propane from the butane, the propane would still have too much volatile ethane and methane in it to allow it to meet market specifications for propane.

To remove the ethane and lighter, the stream goes to an absorber, a column with perforated trays or one filled with some loose packing material. The propane/butane stream (with the little bit of methane and ethane) goes into the bottom of the odd-shaped column as a vapor by letting the pressure down. Further up the column goes an absorption stream called *lean oil* as a liquid. Typically, heptane (C_7H_{16}) is used. It can absorb the propane and butane, but the ethane and methane are too volatile to be absorbed. The lean oil and propane/butane streams slosh past each other through the internals of the column as gravity pulls the lean oil down and lets the propane/butane vapor go up. The lean oil absorbs the propane and butane as it passes by. The ethane and methane bubble up and continue out the top of the column.

From the bottom of the column comes *fat oil,* the lean oil/propane/butane mix, but no ethane and methane. The fat oil goes to a fractionator to strip out the propane/butane. The boiling points of heptane (98.4°F) and butane (32.2°F) are far enough apart that the split is not too difficult, not like trying to split ethane (–132 F) and lighter from propane (–43.8°F). And that is why an absorber system is cheaper to use than a fractionator—smaller columns with fewer trays and less refrigeration or compression.

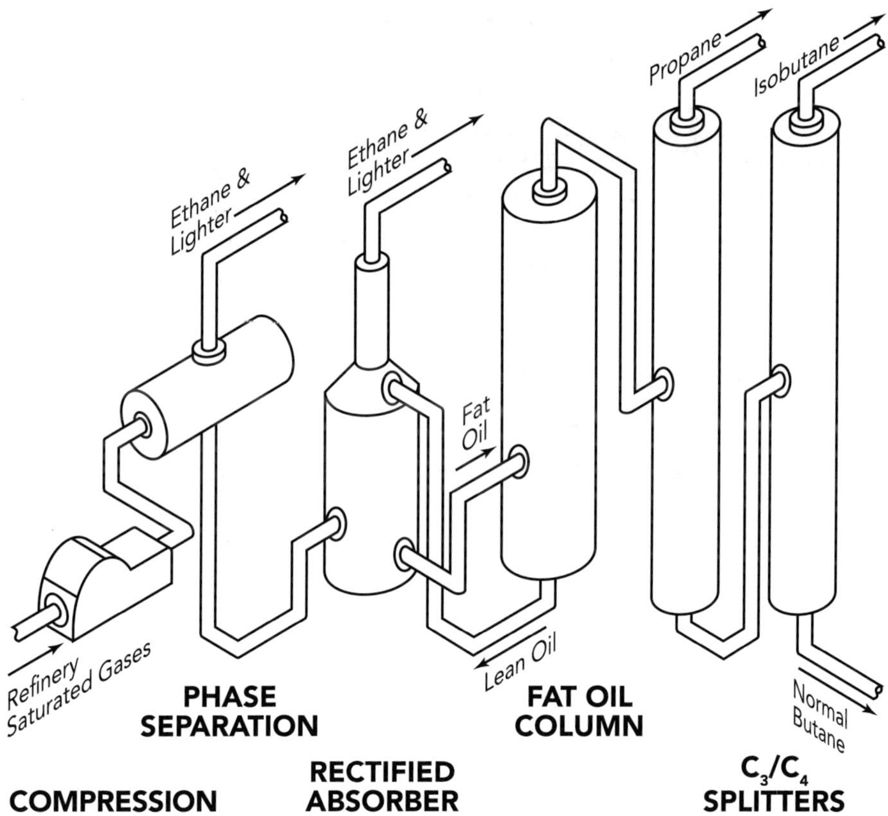

Fig. 6–6. Refinery gas plant: separation by compression and absorption

Treating

Crude oil comes into a refinery with different kinds of contaminants—particularly sulfur compounds. Environmental regulations in most places mandate the removal of almost all sulfur in any form from most of the products in most places. To do this, refineries use several different kinds of treating units, mostly hydrotreating units. This process uses heat, pressure, and catalysts to force any sulfur compounds to react with the abundance of hydrogen pumped into these units. Typically, the sulfur atom is attached to a hydrocarbon molecule, sometimes being deeply embedded in it. As in figure 6–7, when the hydrogen attacks the sulfur, the hydrocarbon molecule breaks down. The desired result is H_2S and a larger, sulfur-free molecule. In the process, by-product C_4's and lighter get created. The H_2S goes to a sulfur recovery plant where it is converted to pure sulfur. The C_4's and lighter end up going to the refinery gas plant for separation.

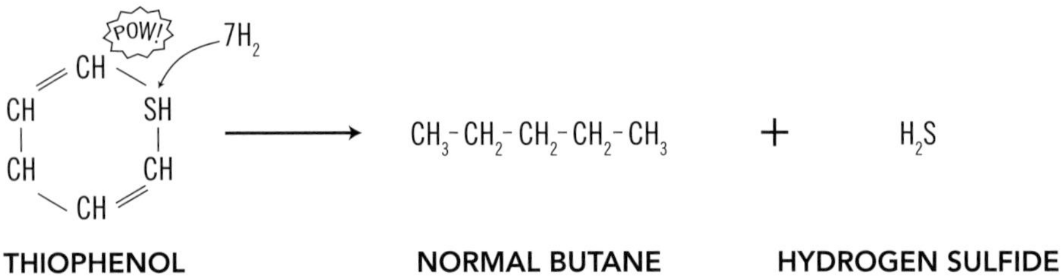

Fig. 6–7. Chemistry of desulfurization

Product Blending

More than a score of different hydrocarbon streams move around a refinery. To generate marketable products, groups of them have to be blended together in a way that they meet market and regulatory specifications. Motor gasoline, diesel fuel, home heating oil, and industrial fuel oil, like many mediocre California wines, are blends of various components. Not exactly so for the NGLs. The C_3 coming out of the refinery gas plant and the alkylation plant as "purity propane" goes right into the marketplace. Normal butane coming into and that made in the refinery is used as a blending component to make motor gasoline. So is the natural gasoline that comes in. Most refineries need more normal butane to produce motor gasoline than they produce themselves. Fortunately, the natural gas–processing sector makes available ample supplies of that commodity.

What used to be decades ago a simple mixing job of the refinery streams and butanes and natural gasoline to make motor gasoline has now been made so complicated by both automotive technology and environmental regulation that it warrants the more elaborate treatment in chapter 8, on NGL markets.

Refinery Outturns of Gas Liquids

Every refinery has a unique assembly of operating units of different sizes as well as a different and sometimes endlessly changing slate of crude oil sources. Table 6–1 shows the gas liquids from a typical—but not average—refinery, more or less like the one in figure 8–1, which takes in 100,000 barrels per day of diluted crude (butanes have been added to the crude in the field.)

Table 6–1. Sources and (uses) of gas liquids in a refinery running 100,000 barrels per day (MB/D) diluted 24°API crude

	MB/D			
	C_2	C_3	nC_4	iC_4
Distilling	—	100	3,200	800
Cat reforming	100	200	400	400
Isomerization	50	50	50	100
Cat cracking	100	200	200	100
Alkylation	—	—	—	(2,200)
Butane isomerization			(350)	350
Hydrotreating	100	250	400	300
Coker	100	250	250	150
Gasoline blending	—	—	(6,000)	—
Fuel system	(450)	—	—	—
To/(from) outside		1,050	(1,850)	—

Notes:
1. Propane and normal butane from the cat cracker and coker are recovered in the alkylation plant. All others get recovered in the refinery gas plant.
2. A refinery with a hydrocracker instead of a cat cracker may have a smaller alkylation plant to handle only coker gas liquids; also hydrocracking would generate more net gas liquids.

Takeaways from Refineries and the Unnatural Gas Liquids[1]

- Refineries generate gas liquids and consume them as well.
- The process units downstream of the crude unit create streams with different properties as they crack, reform, or otherwise restructure the molecules in the feedstock that goes through them.
- These process units create by-product propane and butanes along the way.
- Refineries make gasoline, diesel fuel, and other products by blending groups of streams from several process units.
- Refineries use their normal butane internally to blend motor gasoline. Gasoline sucks up so much normal butane that many refineries buy more from the gas plant sector.
- Refineries may use isobutane in their alkylation plants to make propylene and butylenes into alkylate, a high-octane gasoline blending component.
- Propane is propane, the only one-molecule product—not a blend.

Exercises

1. How are the separation units fundamentally different from all the other units in a refinery?
2. How do natural gas plants differ from refinery gas plants?

1 If these takeaways go in the opposite direction of your curiosity, you can find much more detail in *Petroleum Refining in Nontechnical Language*, 4th Edition, by William L. Leffler, Tulsa, OK: PennWell, 2008.

CHAPTER 7

Logistics

All movements go too far.

—Bertrand Russell (1872–1970)

In 1973 operators in Kingman, Arizona, detected a leaky fitting during transfer of propane from a rail car to a storage tank. One operator inexplicably attempted to fix the fitting by striking it with a wrench. A spark ignited the accumulated vapors, and the flame raced back to the tank car. In a few minutes, a V-shaped flame shot 80 feet into the air from the tank car fittings, disappeared, shot out again, and disappeared. As the local firefighters sprayed water on the tank car, it exploded in a BLEVE (boiling liquid expanding vapor explosion). Debris flew for 2,000 feet; the tank car settled down 1,200 feet down the track; and the explosion left a 10-foot-deep crater. Eleven firefighters died, and 107 people suffered burn injuries.

Natural gas liquids (NGLs) are moved about by almost every conceivable mode of transportation: big trucks, little trucks, railcars, metal cylinders, pipelines, barges, and ocean-going tankers. When they are not being moved, they are kept in unusual storage facilities, some far beneath the Earth's surface.

Overland

As early as 1912, distributors moved propane and butane, not yet called liquefied petroleum gas (LPG), in metal cylinders, hauled on pickup and flatbed trucks. By 1928, the first bobtail truck, designed to deliver LPG to the retail trade, when into service. That reduced the number of trips needed to keep rural America warm, as more customers placed 500- to 1,000-gallon tanks in their yards, resulting in larger, less frequent deliveries.

Trucks

Two types of trucks haul LPG: the highway transport and the bobtail truck. The first is used when gas plants producing propane and butane have no access to an LPG pipeline, and the product must be moved over land. These plants have large enough storage onsite to load highway transport trucks that have capacities of 9,000–12,000 gallons. These large tankers, sometimes called a "pipeline on wheels," move propane directly to retail distribution centers, large storage facilities, petrochemical plants, or pipeline terminals for further transport.

The smaller size truck, the bobtail, handles retail distribution from the depot. Bobtail tank capacities normally range from 750 to 6,500 gallons, with the most common sizes in the 2,800- to 3,500-gallon range (fig. 7–1).

Fig. 7–1. Propane bobtail truck (courtesy of Jarco)

The size of the bobtail that a retailer uses depends on the onsite storage, the type and location of customers, and their storage and use patterns. For backyard and other awkward deliveries, the bobtails may have long-reach delivery hoses, rewound on reels by electric or air-powered motors.

As in any large bulk-carrying truck or tank car, a partial load of liquid can present a particular hazard when the vehicle is in motion. Sudden turns, stops, or starts cause the liquid to slosh and can affect the stability of the vehicle. An emergency stop of a bobtail, for example, might result in a lurch forward as the liquid surges against the front of the tank. For that reason, the tanks have built-in baffles (fig. 7–2) to reduce the movement of the contents.

Fig. 7–2. Cutaway of a bobtail showing the baffles

Operations. Filling the pressurized tank requires a reduction of the LPG vapors left from the last load. As liquid propane is pumped into the fill line, propane vapor liquefies to make room. It shrinks 60 to 1 in volume in the process. Pumping in liquids raises the pressure in the tank, which compresses the vapor, condensing it to liquid.

Pressures in these trucks are rated at a minimum design pressure of 100 psi and maximum of 500 psi. The trucks have liquid level gauges, but the pressure tells a lot about how much LPG is in the tank. The more liquid, the higher the pressure, and vice versa. This relationship is complicated by temperature: The warmer the day, the higher the pressure, obscuring the exact amount implied by the pressure gauge.

Wholesale deliveries are normally made by large transport vehicles at specially designed unloading stations called bulkheads, or truck risers. Most transports are unloaded by a liquid pump that can transfer as much as 300 gallons per minute of liquid propane. In some locations, compressors may be used. Railroad unloading stations are located on sidings next to the bulk storage facility. Railroad tank cars do not have their own pumping equipment. A compressor is used during liquid transfer to facilitate vapor recovery. An integral part of propane transportation that calls for enhanced driver and loader expertise is the loading and unloading process. Normally, the transfer takes place at a bulk plant, into or out of a storage tank. Often in field locations, transfer directly between a tank car and a tank truck takes place (fig. 7–3).

Fig. 7–3. A propane transport truck unloading from a pressurized tank car (courtesy of Westmor Industries)

Rail tank cars

Moving hydrocarbons by rail evolved slowly after the oil industry started in Titusville, Pennsylvania, in 1859. Wooden barrels on horse-drawn wagons became the transport mode of choice for the first five years, until the railroad realized visions of untold volumes of freight. They experimented with placing wooden staved vessels holding 400 barrels on flatcars for a while. Also, multitiered 50-gallon barrel carriers combined economies of scale with ease of redistribution at the destination.

Propane and butane represented a more technical challenge because of their volatility. Tank cars capable of handling the pressures to keep these fuels in the liquid state appeared on the rails as early around 1922 after tank car manufacturers started using welding instead of riveting. The car capacities were generally 8,000–10,000 gallons of liquid. In the 1960s car manufacturers began producing 30,000- to 32,000-gallon cars, then called jumbos. Railroads accommodatingly changed their tariffs to set the rates per car so that the cost per gallon of propane or butane dropped by about 30%. For a while, the railroads offered to haul three 10,000-gallon cars locked together for the price of a jumbo. Eventually, nearly all the small cars exited the business, and the standard parcel size settled at the jumbo's 32,000-gallon tank car (fig. 7–4).

Fig. 7–4. Standard 32,500-gallon pressurized tank car (courtesy of American Railcar Industries)

In the United States, most companies lease their tank cars from the big tank car leasing companies (Union Tank Car Company, GATX, and others). Railroads gave up supplying tank cars a long time ago. Still when a shipper provides his own car (or a lease car), a railroad provides a rebate because it doesn't (or never did) provide the car.

Construction. Design specifications for propane and butane tank cars changed after a calamitous incident in Kingman, Arizona, in 1973. The U.S. Department of Transportation issued design regulations that required that tank cars hauling flammable materials have outside the steel shell a layer of thermal insulation, covered by a layer of steel that holds the insulation in place but does not add to the pressure-containing ability. The insulation protects the tank car in a 1,600°F fire for 100 minutes or in a 2,200°F fire. That gives the firefighters time to get cooling water on the tank car or to evacuate the area if they fear an imminent BLEVE (boiling liquid expanding vapor explosion; see chapter 10 for more information about BLEVEs.)

Since the 1960s, design specifications call for railcars to have top loading and unloading, a safety feature to eliminate impact accidents. The top dome (fig. 7–5) has the fittings, including vapor venting. Tank cars have other safety devices, including pressure relief valves and heavy-duty steel head shields to prevent puncture during incidents.

Fig. 7–5. Fittings in the dome of a propane tank car

Underground

All the NGLs can move by pipeline. Moving pressurized gas by pipeline started well before the 1859 Titusville, Pennsylvania, oil discovery. The London and Westminster Light and Coke Company began laying the first gas mains ever placed under a public street in 1806. They were supplying gas manufactured from coal to the streetlights of London. The practice jumped to U.S. cities starting in 1825. Remnants of gas light districts still survive in several cities.

Pipelining the newfound Pennsylvania crude oil started with fits in the 1860s as producers attempted to eliminate hauling oil barrels for just a few miles from wellhead to refinery. In 1879, the Tidewater Pipeline Company completed the first crude oil trunk line from western Pennsylvania to Williamsport, Pennsylvania, a 115-mile trip.

All along, manufacturers made pipe by taking iron plate, rolling it into a tube, and welding the seam. In 1885, the Mannesmann brothers, Rinehart and Max, of Germany devised a machine that could make seamless pipe by pushing a mandrel (a big shaft) through a rotating billet of hot steel. Seamless pipe enables much higher pressures within an operating pipeline. Large gangs were needed to screw together threaded lengths of pipe until the 1910s, when the welding of joints started.

In 1941 the Shamrock Oil Company finally got the engineering temerity to construct long-distance NGL pipelines, which required much higher pressures. Together with Phillips Petroleum, they built pipelines from the Texas Panhandle, the major source of NGLs at the time, to destinations in south and east Texas, hundreds of miles away.

Movement

The NGL cargo in pipelines can be transported in several ways (fig. 7–6): Y-grade (or raw make) lines that hold a mix of ethane through natural gasoline; dedicated lines for single NGLs, mostly ethane, propane, or butane; batch lines, where propane or butane are interspersed between parcels of crude oil or products; and diluted crude lines, where butane or natural gasoline are admixed with crude oil as an economic way to transport them.

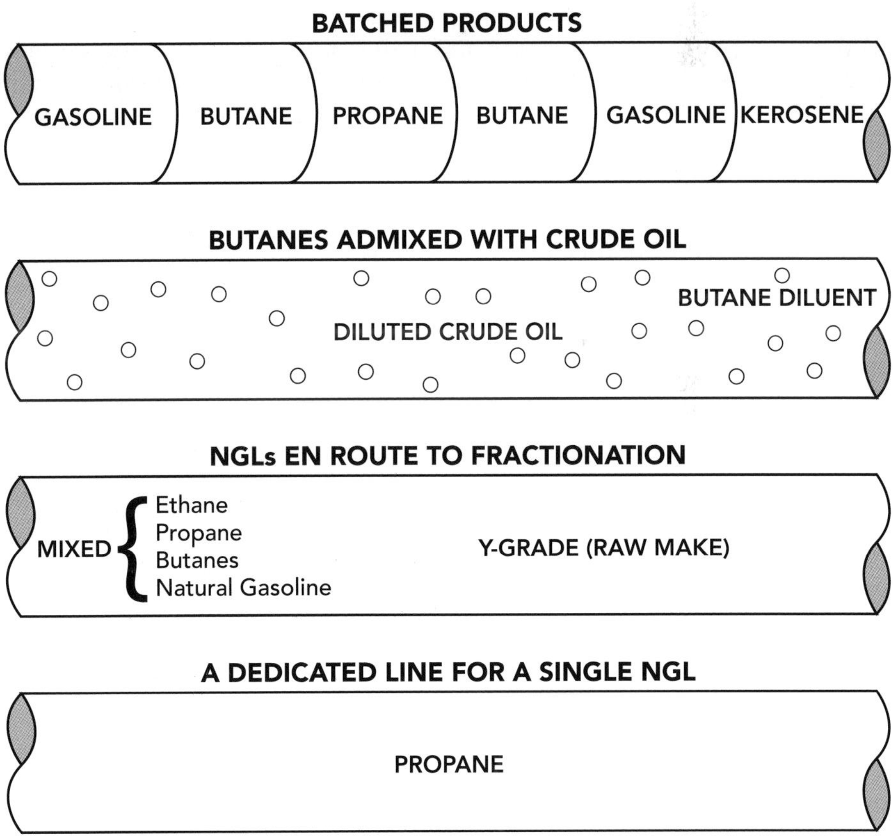

Fig. 7–6. Moving NGLs in a pipeline

Single NGL lines

Propane and butane pipelines operate at 100 to about 1,000 psi to keep the product in the liquid state, an easier way to transport and store it. Ethane is also usually handled that way, but this can be complicated by its critical temperature. This arcane concept comes from a remote corner of thermodynamics and vexes ethane pipeline operators in those parts of the world where ambient temperatures sometimes run well over 100°F. In those places ethane in a pipeline can pick up enough heat to exceed its *critical temperature*, 90°F. At that point, no matter how high the pipeline pressure reaches, the ethane will not liquefy. It will be a *supercritical fluid*. Increasing the pressure will reduce the volume of the vapor content some, but will not eliminate it by liquefying it. (Ethylene is even more problematical. It has a critical temperature of only 48.5°F, so most ethylene lines have supercritical ethylene in them all the time.)

Temperatures below critical can present more problematic situations. At some combinations of pressure and temperature below the critical point (critical temperature and critical pressure), the fluid may consist of both liquid and vapor. Pipeline conditions are usually set to operate above the pressure where this happens, so the ethane is either all liquid or all vapor. A diagram showing the vapor/liquid and pressure/temperature relationship is intimidatingly hairy enough that the full description of this strange phenomenon has been deferred to appendix 5. The critical temperatures of propane (616°F) and butanes (~551°F) are so high that the multiphase problem never arises.

Batching

The NGLs, singly or in combination, can move down a multiproduct pipeline in discrete parcels. Counterintuitively, very little product sloshes into an adjoining batch of another product. Each batch just moves down the line nearly intact. Even so, to ensure the integrity of each batch, similar products usually buffer each other so that traces of one in the other do not matter. (Just a little bit of propane in diesel fuel would be really bad, but a little butane in gasoline or propane doesn't matter much.) So in batching NGLs, propane would be buffered on either side with butane, which might be followed by gasoline.

Many pipelines that batch propane use pigs (fig. 7–7) at either end of the parcel. The suction-cup-like ends of the pigs provide a seal to keep one product from leaching into another.

Pigs?

How did the devices used to separate products, to clean pipelines, and now to detect mechanical flaws of the lines get the name *pig*? An expert in the field, Andrew Marwood of the Pipeline Engineering Company, offers two possibilities:

- PIG, a pipeline injected gadget
- In the early days of pipelining, as the leather balls used to swab pipelines moved down the line, they squealed—like a pig.

Marwood suggests that both undocumented possibilities may be may be more a myth of mirth than a matter of fact.

Fig. 7–7. Pipeline pig

Diluted crude

Many crude oils have very low vapor pressure, well below the limits of the pipeline they move in. Crude oil can absorb butanes like a sponge (within limits, of course.) Refiners are usually happy to receive this inexpensively moved source of butane and separate it from the crude oil in their crude oil distilling unit.

Natural gasoline, of course, creates no vapor pressure problem either. Pipelines do allow natural gasoline to be admixed with crude oil. In fact, for very heavy crude oils such as those in western Canada, diluting them with natural gasoline or condensate is the accepted way to reduce the viscosity and make the heavy crude pumpable.

Injecting butane or natural gasoline into a crude oil line alters the value of the material being pipelined. For that reason, tedious bookkeeping keeps track of these mixtures of multipriced commodities so that the producers receive fair compensation and refineries pay for what they get.

Seaborne

Propane and butanes have moved across the seas for decades. At first, the technique was unsurprisingly only an extension of overland transportation: Put steel cylinders on board and fill them up. Eventually, purpose-built vessels capable of holding hundreds of thousands of barrels plied the sea lanes. River trade by barge, a smaller volume but still important activity, followed the same path.

Tankers

Until the post–World War II period, NGLs remained multilocal markets—lots of markets around the world that were unconnected by transportation. That changed in 1947, when Warren Petroleum commissioned the LPG tanker, the *Natalie Warren,* and began moving product from the Esso refinery in Aruba to Brazil. They refitted a dry cargo freighter by installing 68 pressure vessels in five cargo holds to handle 100,000 gallons. Esso quickly followed with a small fleet of its own.

Since the design pressure of the steel tanks was 240 psi, the tanks themselves weighed as much as the cargo. In the ensuing decades, engineering efforts responded to the demands of much longer hauls—exports from the Middle East and Venezuela, and imports by companies in the Far East, Europe, and the United States. First came semirefrigerated ships, with plants on board that could keep the cargo at low enough temperatures to cut in half the pressure necessary to keep the cargo in a liquid state. That represented a quantum leap in the carrying capacity of new builds to 1–1.5 million gallons.

In 1962 the Mitsubishi shipyard built the *Bridgestone Maru,* another new class of LPG ship, the very large gas carrier (VLGC). The refrigeration plants on these huge vessels kept the cargo at –45°F, below propane's boiling temperature. That allowed virtually zero working pressure and less tank weight (but more expensive nickel-steel alloy to deal with the low temperatures.) The tanks no longer had to be spherical and could match the contours of the ship. VLGCs now carry more than 20 million gallons of LPG (fig. 7–8).

Refrigeration facilities at the origin prepare the LPG for loading. At the destination, refrigerated tanks receive the supercooled cargo; heaters raise the temperature before the product moves into commerce.

Fig. 7–8. VLGC *Mill Reef* (courtesy of Petredec)

Storage

All hydrocarbons have to be stored at places along their value chain. NGLs present three different sets of requirements and three different options, mostly matching one to one.

Bullets, spheres, and tanks and cylinders

Working storage at gas plants and refineries for natural gasoline, the heaviest of the NGLs, can be like storage for most other oil products—large cylindrical tanks. Currently, most of the time (though not always), natural gasoline, like finished motor gasoline, has to be stored in tanks with a closed system and vapor recovery to ensure that vapors do not reach the environment.

For ethane, gas plants and/or petrochemical plants have steel surge tanks to level out intermittent variations in production, shipping, or consumption. Storing more than a few hours worth of ethane movements in steel tanks is not common, and the tanks have to be able to handle pressures in the hundreds of psi.

Propane and butane have yet another requirement for working storage. These two NGLs need inventory accumulation that can last for days or months—days where the offtakes move out more or less steadily, but at the mercy of truck and tank car arrivals and pipeline space; months where they provide fuel for increased seasonal heating demands. Typically above ground, propane or butane steel storage has the characteristic rounded structure, either spheres or cylinders with rounded ends called bullets (fig. 7–9). Internal pressure in these vessels can reach hundreds of psi, especially during hot weather. The rounded structure is engineered to spread the internal pressure to reduce the tank's structural steel requirement. That makes it cheaper to build.

Fig. 7–9. Steel spheres for propane storage (courtesy of Dominion)

Some locations use refrigeration to reduce steel requirements. By cooling the propane to –41°F, which is below its boiling point, or butane down to 9°F, which is below the isobutane boiling point, the material can be stored in a tank at ambient pressures in the tank. Of course, the installation requires a reliable refrigeration plant, an insulated tank to reduce heat absorption from the environment, and safety components to flare any emergencies when the temperature might rise.

At depots, industrial sites, and residential locations, the common storage facility is the bullet. Depots and industrial sites might have tanks handling 100,000 gallons or more. Transfer rates from trucks or tank cars to these tanks run about 300 gallons per minute.

For residential storage, a home might have a 500- to 1,000-gallon tank in the backyard providing heating and cooking fuel supply. The smaller tanks for propane-fired patio grills generally hold about 5 gallons. When the small tank is depleted, the owner needs to exchange it for a full one. Empties are usually exchangeable at depots, hardware stores, and even drugstores (fig. 7–10). Propane distributors can refill them at their depots. Both steel and fiberglass composite tanks are now on the market.

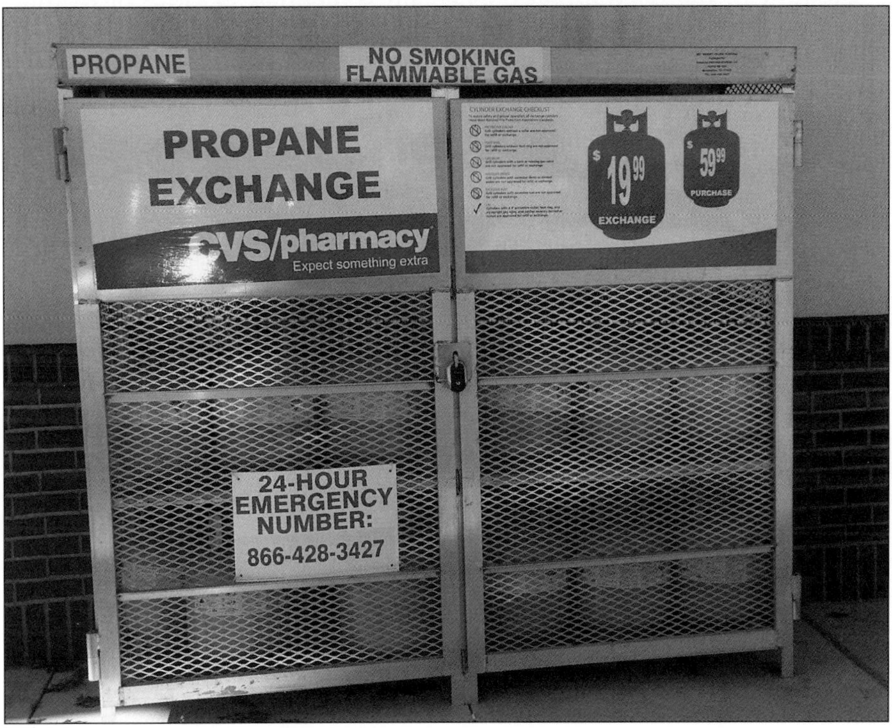

Fig. 7–10. Ubiquitous propane cylinder exchange cage

Each bulk plant is different. However, typical equipment includes bulk storage tanks, unloading stations, and filling stations. Bulk plants may have separate facilities for filling cylinders; using smaller pumps with a flow rate of 10–15 gallons per minute is common. Large numbers of cylinders, either filled or empty, may be warehoused at these bulk plants.

Underground storage

Salt. A long, long time ago, landlocked saltwater covered large areas of the Earth. In a prime example, a land bridge from what is now the Yucatan Peninsula through Cuba, the Caymans, and surrounding islands to the Florida peninsula enclosed the Gulf of Mexico. Apparently, ancient rivers from North America were insufficient to keep the gulf full, and it began to evaporate. As the saline concentration increased, large volumes of salt precipitated out of the evaporating water, leaving huge salt layers at the sea bottom. Over ensuing eons, sediment washing down from the continent covered the salt with thousands of feet of overburden, actually moving the shoreline out hundreds of miles.

Fissures in the overburden opened up sometimes because of tectonic plate movement or from other causes. Large masses of salt are startlingly pliable and mobile, and slowly—very slowly—over the eons, the pressure of the overburden squeezed the salt into the fissures. Huge masses of salt moved toward the surface, not unlike the blobs in a lava lamp, and formed subsurface salt domes of various shapes. They have taken the shapes of columns, elongated overhangs (fig. 7–11), and even ice cream cones.

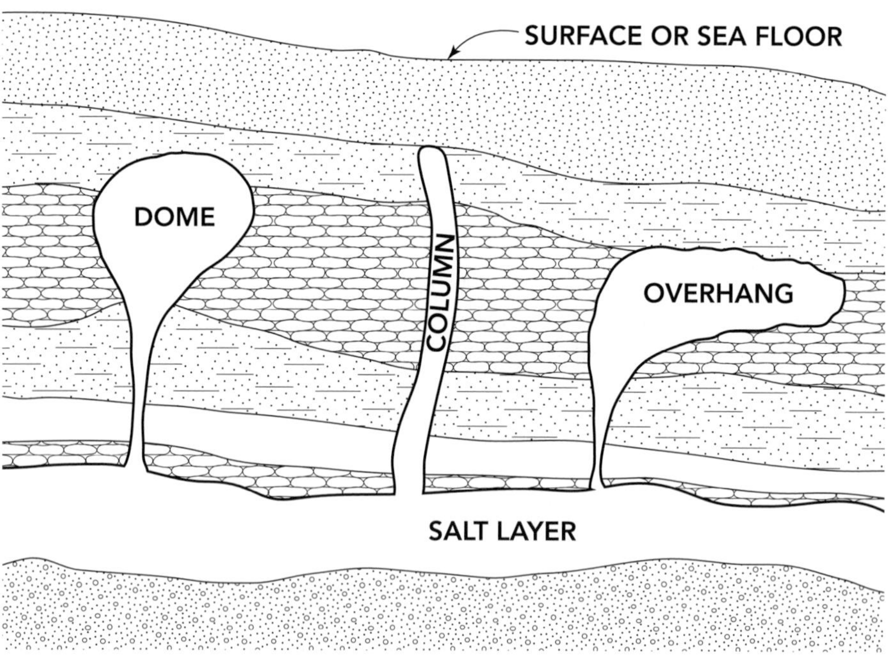

Fig. 7–11. Salt dome formations

Numerous other locations around the world, many of those curiously located inland, have these thick salt deposits. As difficult as it is for the hardscrabble farmers there to believe it, even West Texas and Oklahoma were once covered by a saltwater sea. As in chapter 2, once Smokey Billue established in 1950 that cavities in some salt deposits can hold hydrocarbons, the rush to store LPG and other petroleum products below the ground began.

Engineering. To create a salt cavern or storage jug, after the engineering surveillance maps the shape and size of the salt accumulation, a hole is drilled down several thousand feet into the salt layer. Two sets of steel casing are placed in the hole and cemented in place. Freshwater is pumped into one. The water dissolves the salt and returns it to the surface via the other pipe as brine (saltwater.) The brine moves into a brine pit nearby. The freshwater pumping continues until enough salt has dissolved to give the size cavity needed. Cavern capacities vary from about 100,000 barrels to more than 500,000 barrels.

Once in operation, one of the pipes into the jug is hooked up to a propane (or whatever liquid needs storing) line. As propane is pumped into the jug, it floats atop the water, as all hydrocarbons do, but forces brine out from the bottom (fig. 7–12). To recover the propane when it is demanded, water is pumped from the brine pit back into the jug, forcing the propane out. The propane goes through a dryer to remove the water vapor that inevitably drifts in.

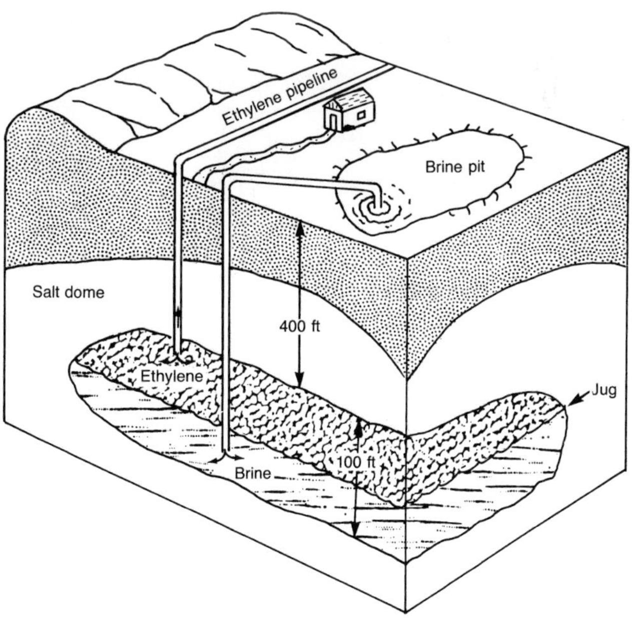

Fig. 7–12. Salt dome storage cavern

Salt caverns accommodate many different commodities: NGLs, crude oil, and petrochemicals such as ethylene and propylene. Petroleum products such as gasoline and diesel fuel are generally not stored in salt caverns. They are more like beer than wine—they get worse with age. Because of the cracking processes in refineries, they contain olefins, those double-bonded, reactive molecules that can react with each other, forming gums and other bad actors over time. (That's the same reason why it's not good to leave gasoline in your lawnmower over the winter—it gums up the spark plug and fuel injector.)

Rock. Some places have no salt, but they have rock, and some of it is so tight that caverns can be hewn in them to contain NGLs or other commodities (fig. 7–13). Suitable underground formations include granite, limestone, and shale. The caverns are unlined, and most depend on the water that saturates the subsurface where these caverns are built. The pressure from the water in the rock, high due to the hundreds of feet of sediment above, keeps any hydrocarbons from slowly leaching into the rock formation.

Fig. 7–13. Rock cavern ready to be filled (courtesy of Geostock US)

Rock caverns are mined starting with a borehole wide enough to drop mining equipment down a thousand or more feet. After an initial widening of the cavern at the bottom, more equipment is lowered, allowing full-scale mining, but evacuation through a minimalist access is a tedious process. At completion, the equipment is withdrawn, and the access hole is sealed except for the cemented casing used to fill and withdraw the product. Just as with steel storage, in operation the cavern space will contain with both liquid propane, for example, and propane vapor on top. Rock caverns are always deep enough that they maintain a constant temperature year round, making measurement a bit easier than aboveground storage. Some rock caverns can store more than a million barrels.

A few rock caverns originated as metal or coal mines, and they were subsequently carefully fitted with casing and sealed off.

Takeaways from Logistics[1]

- Ethane moves almost exclusively by pipeline, though a bit moves by sea in tankers.
- Propane and butane move as liquids in pressurized pipelines, railcars, trucks, barges, and tankers.
- NGLs can be stored in large volumes in salt caverns or rock caverns. Small volumes can be handled in pressurized steel tanks or refrigerated tanks.

Exercises

1. What are five ways a refinery can receive a supply of butanes?
2. How can a refinery get a more assured supply of isobutane?
3. Which NGLs are shipped as liquids and which as gas?
4. What are the cheapest ways to store NGLs?

1 If your curiosity is now piqued, you can find much more detail in *Oil & Gas Pipelines in Nontechnical Language*, by Thomas O. Miesner and William L. Leffler. Tulsa, OK: PennWell, 2006.

CHAPTER 8

NGL Markets—Petrochemicals

Let us break the bonds asunder: and cast away the cords from us.

—Prayer Book of the Church of England, 1662

To what degree?

Neanderthals and *Homo sapiens*, lager and ale, Fahrenheit and Celsius—why are there always two kinds of everything? Well, in 1724 Daniel Fahrenheit, a brewer keenly interested in measuring temperature, created a scale based on three points. He set the temperature of the coldest liquid known at the time, ice water with ammonium chloride, at 0°F; plain ice water at 32°F; and the human body temperature at 96°F (he thought). By experimentation, he then observed water boiling at 212°F.

"Too awkward," said Anders Celsius, who in 1742 proposed a more orderly scale with boiling water at 0°C and ice water at 100°C. Right after Celsius died, Carl Linnaeus, without totally clear foresight, thought that people would always be interested in hotter temperatures, not colder, so he reversed the scale—ice water at 0°C and steam at 100°C. Preoccupied with hot things, he thought that would minimize having to use so many negative numbers. (He had not heard about the notion of cryogenics.) Linnaeus was kindhearted enough to leave the scale's name unchanged.

Today, Europeans and scientists favor Celsius. Most Americans use Fahrenheit.

Petrochemical plants have a teenager's appetite for all the natural gas liquids (NGLs), especially ethane and propane. Some parts of the world have little or no natural gas or NGLs, so olefin plant operators turned to naphthas. Plant operators buy them from refineries or use natural gasoline or butane available by tanker from other parts of the world.

Olefin Plants

Chapter 4 explained the libertine nature of ethylene and propylene. They seem to want to conjoin with almost every other kind of molecule—or each other. For that reason, the two are the building blocks for most of the commercial petrochemical products produced today. Their double bonds make them ready to react with themselves (to make polyethylene and polypropylene), other building blocks such as butylene (linear low-density polyethylene), benzene (styrene and polystyrene), atoms such as chlorine (vinyl chloride and polyvinylchloride), and many others.

Making ethylene and propylene from NGLs in olefin plants (crackers) does not require complicated chemistry. Figure 8–1 shows some of the variation that occurs from cracking ethane and propane. At the top, ethane cracks into ethylene and two hydrogen molecules. Further down, bonds can sever to create, regrettably, two methane molecules. Even a little acetylene can result from cracking ethane.

Cracking propane can yield ethylene or propylene or, not shown, some other by-product molecules of not much interest to the petrochemical manufacturers.

Cracking butane or natural gasoline has more complicated results. Both are bigger molecules to start with, especially natural gasoline, which has five, six, seven, and maybe even eight or nine carbon atoms, with all sorts of shapes and isomers. As a consequence, the yields shown in table 8–1 have more by-products besides ethylene and propylene.

Table 8–1 shows that the more complicated the feed, the lower the yield of ethylene and the higher the yield of propylene and by-products. Not that this is all bad. The butane/butylenes have a ready home in refinery alkylation plants (as in chapter 6) and petrochemical plants. Butadiene is used to make rubber and other elastomers, plastics that stretch and bend easily. Refineries buy the pyrolysis gasoline that comes from cracking naphtha/natural gasoline to blend into motor gasoline after treatment. The methane/hydrogen stream, which is mostly methane, is used for plant fuel and is really the only loser—the value of this stream as fuel is lower than the cost of the raw material feed to the plant.

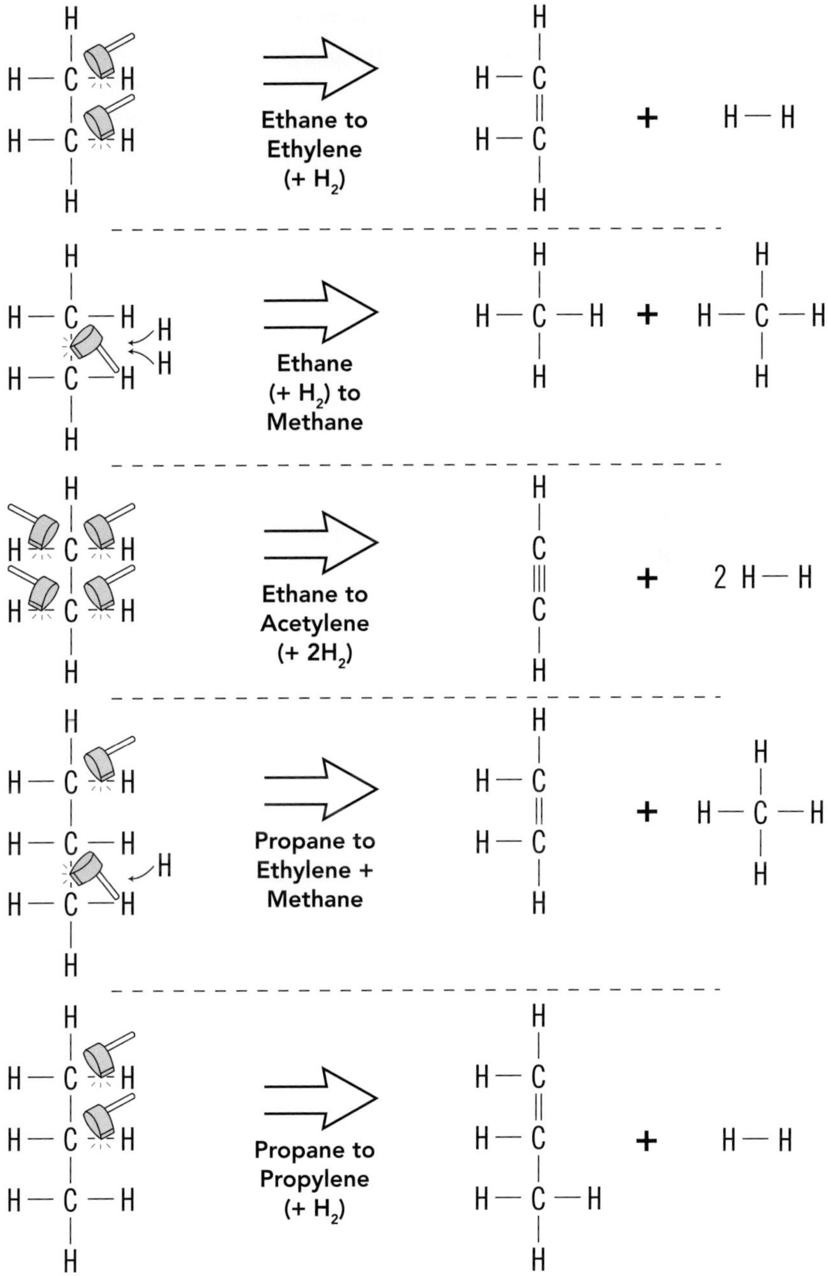

Fig. 8–1. Cracking: Ethane and propane crack into ethylene, propylene, and by-products.

Table 8–1. Olefin plant yields from NGLs (in pounds of product per pound of feed)

	Ethane	Propane	Butane	Natural Gasoline and Naphtha
Ethylene	0.80	0.40	0.36	0.23
Propane/propylene	0.03	0.08	0.20	0.13
Butane/butylene	0.02	0.02	0.05	0.15
Butadiene	0.01	0.01	0.03	0.04
Fuel gas	0.13	0.38	0.30	0.26
Pygas	0.01	0.01	0.06	0.19

The olefin plant

The transformation of the NGLs into olefins takes place in an olefin plant, sometimes called an ethylene plant, sometimes an ethane or propane or liquids cracker, and sometimes an ethylene cracker (a misnomer, because the ethylene is not cracked.) The simplest example of any of them is an ethane cracker, shown in figure 8–2. This is one of the few petrochemical processes that do not use a catalyst to promote the reaction. The cracking happens as the feed, ethane in this case, is heated to about 1,600°F in a furnace. (The ethane stays inside the tubes in the furnace and does not come in contact with the furnace flames. Otherwise there would be a really snappy reaction.) The ethane stays in this pyrolysis section (the Greek word pyro means fire) for just a couple tenths of a second. At that temperature the ethane gets so agitated from all the extra energy that it cracks—severs the bonds in one or more places.

Coming out of the pyrolysis furnaces, the cracked products need to be cooled down right away. Otherwise they keep cracking all the way to methane. The stream then goes to a *quench pot* to the drop the temperature by 800°F–1,000°F.

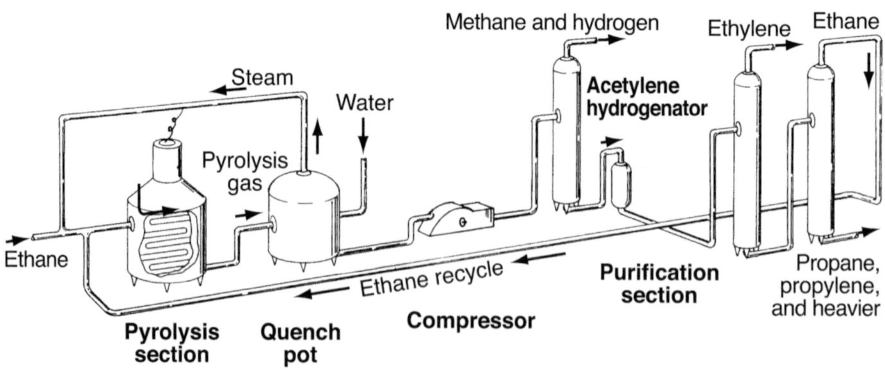

Fig. 8–2. An ethane cracker

Out of the pyrolysis section comes ethylene, by-products, and about 20% ethane since not all the ethane cracks in one pass. The cooled stream goes through the separation section: compression to liquefy most of it, then to several fractionating columns to separate the hydrogen/methane, the ethylene, the ethane, and the heavier by-products. The ethane is recycled for another pass at cracking. In fact, no ethane leaves the plant; it is recycled to extinction.

An ethane cracker is relatively simple because the pyrolysis furnace generates only a few things to be separated. At the other end of the NGLs spectrum, cracking natural gasoline in a naphtha cracker creates a more extensive, heavier array of streams. For that reason, the furnace is a larger and the downstream separation section has more fractionating columns. The natural gasoline/naphtha cracker in figure 8–3 has an alternative piece of equipment called a *cold box* between the pyrolysis section and the separation section. Keeping in mind that the feed to a distilling column has to be part liquid and part vapor, then instead of compression (high pressure) to turn most of the cracked stream to liquid, a powerful refrigerator that drops the temperature does the same thing.

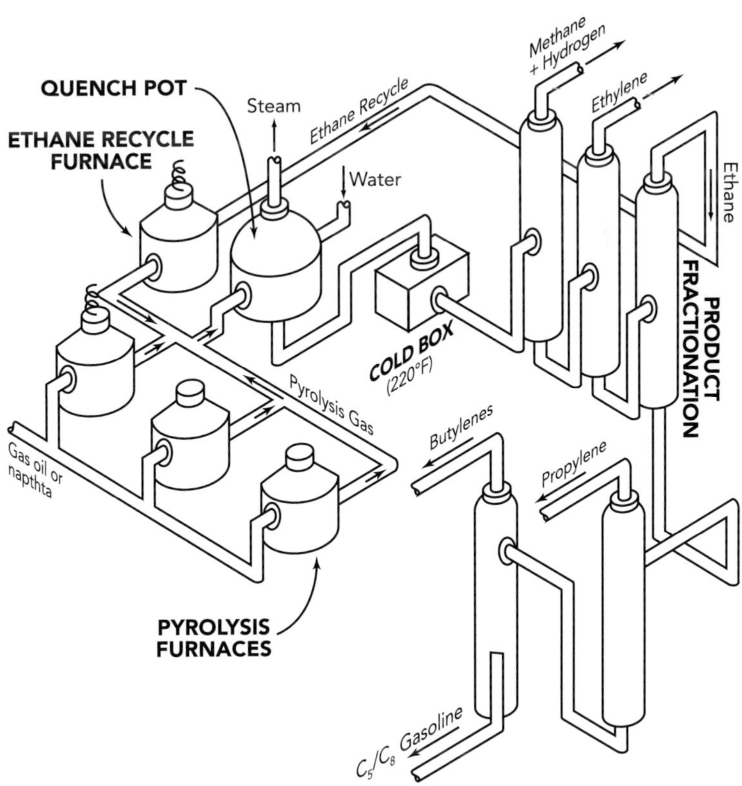

Fig. 8–3. An olefin plant that can crack natural gasoline/naphtha and lighter NGLs

Naphtha crackers (fig. 8–3) also create an ethane by-product so they have an ethane furnace. Again, the ethane is cut out in a column in the separation section for recycle.

Most naphtha crackers can also handle the other NGLs. The separation section has to be run with care. Some separation columns handling the heavy products might have to be shut down when running ethane or propane because these lighter NGLs have such low yields of heavy ends.

Propylene grades

The propylene stream coming from cracking propane, butane, or naphtha usually has 6%–10% propane along with it and is called *chemical grade propylene.* Many petrochemical processes can use propylene of this concentration. Some processes, especially those making polymers (plastics), call for a higher purity propylene, 99.5%–99.9%, called, understandably, *polymer grade propylene.* Since none of the olefin plants produce a propylene stream of this purity, the chemical grade propylene stream can go to a C_3 fractionator (or splitter) to separate the two, propylene and propane, from each other. (The same thing is true for propylene coming from a refinery cracker or coker, called *refinery grade propylene.*) The propane coming from these C_3 splitters goes into the propane market.

Choosing feeds

So with all these feeds available to the olefin plant operator, which one is best? It depends on quite a few factors. Take the actual prices for the NGLs over the 470-day period shown in figure 8–4. By quick observation, someone might be tempted to pick ethane as the best feedstock. Cheapest maybe, but not necessarily the most economical. The thought process of deciding which to buy (fig. 8–5) shows the logic. It takes into account the by-product credits from cracking any of the four feedstock options, in this case natural gasoline.

The yields in the by-product credits box come from Table 8–1, under naphtha/natural gasoline. The by-product prices and the feedstock prices come from the daily market values, as shown in figure 8–4. The feedstock cost of 1 pound of natural gasoline on the one day in the example was 21 cents per pound. The by-product prices were taken from the same day, of course. From Table 8–1 and the by-product prices come the total by-product credits of 5.3 cents per pound. So the net feed (1.0 – 0.64) cost is 21.0 – 5.3 = 15.7 cents per pound. Now that makes only 0.36 pound of ethylene and propylene, so the cost per pound of ethylene and propylene is the 15.7 cents per pound divided by the 0.36 pound produced per pound of feed.

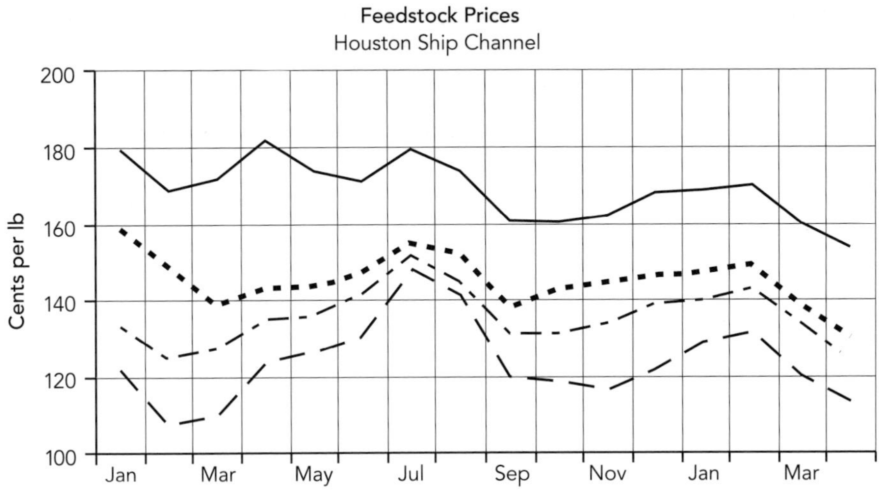

Fig. 8–4. NGL prices along the Houston Ship Channel

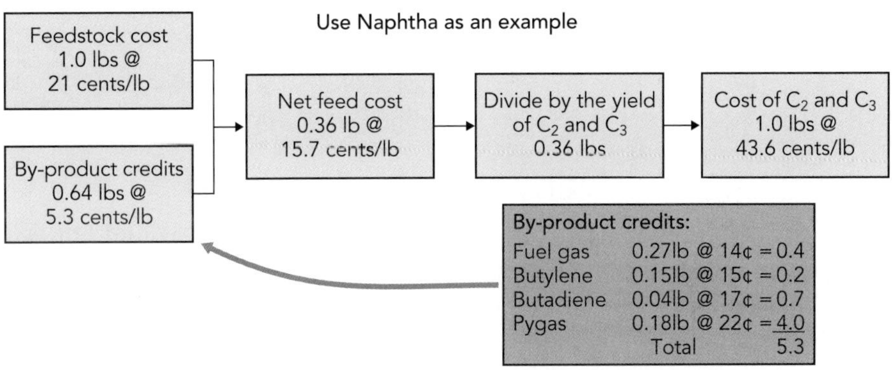

Fig. 8–5. Choosing feeds: calculating the cost of ethylene and propylene

That's just one feed for one day! Do the same calculation for each feed for each of the 470 days on the chart with their changing prices gives the chart in figure 8–6.

During this 15-month period, it turns out the cheapest feedstock, ethane, was not the *most economical* feedstock because of the high values of the by-products from cracking the other feeds. The heavier feeds, butane and sometimes natural gasoline, proved best.

All that depended, of course, on whether the olefin plant had the facilities to handle the heavier feeds. And keep in mind that the market price relations of the NGLs *and* the by-products continuously change over time, making future market price assessment and this calculation a fulltime job of the traders in the back rooms of the olefin company.

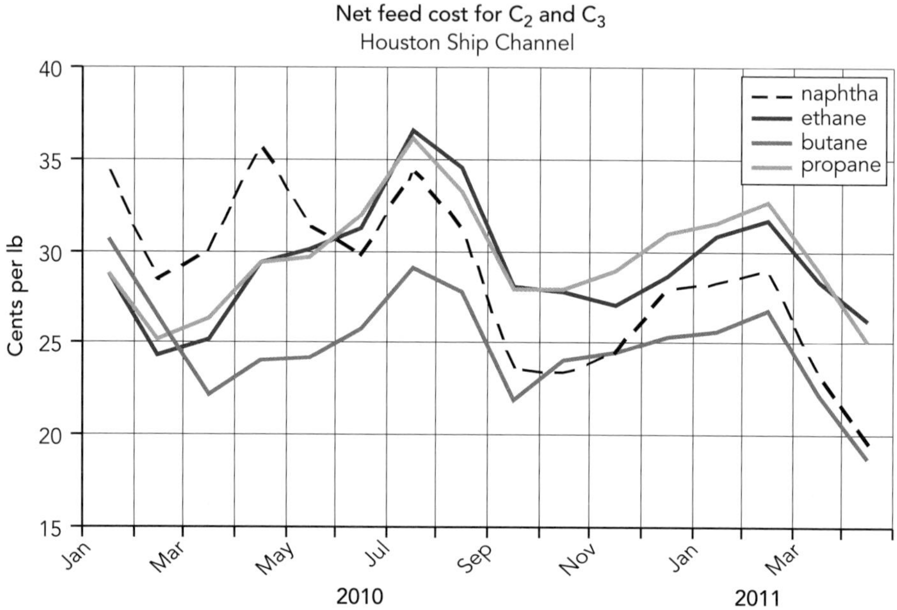

Fig. 8–6. Ethylene and propylene cost calculation results

Dehydrogenation

At various times and places there is plenty of ethylene around but not enough propylene coming out of the olefin plants. That happens when propylene demand grows faster than ethylene demand or when the dominant feedstock is ethane, yielding little else but ethylene.

Rising to the occasion, the chemists and engineers in labs came up with *propane dehydrogenation* (PDH). Broken down, the word is not so intimidating. The difference between propane and propylene is two attached hydrogen atoms. The process hydrogenation adds hydrogen; *de*hydrogenation takes it away. Propane dominates the dehydrogenation world, but some petrochemical plants use that technology also to turn isobutane into isobutylene and butane or butylene into butadiene.

Propane to propylene

As simple as this chemistry is, PDH facilities require tricky handling of the catalyst and the hydrocarbon. Several technology companies license versions of this process. The one in figure 8–7 uses a chromium oxide catalyst stacked in batches in a reactor.

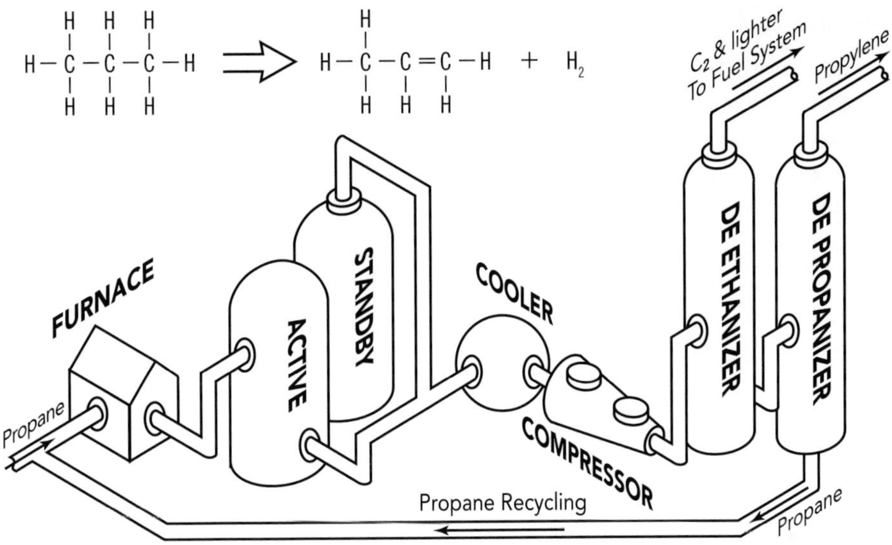

Fig. 8–7. Propane dehydrogenation

About 50% of the propane converts to propylene, plus small amounts hydrogen, methane, and trace amounts of other by-products. Downstream, in the separation section, the unreacted propane is recycled to the reactor.

As the propane flows through shedding hydrogen atoms, occasionally carbon atoms inadvertently pop off the propane molecule and settle on the catalyst. After a while, the carbon covers up the catalyst, which makes it ineffective. At that point, the operators have to switch the propane feed to a second, identical reactor so they can regenerate the catalyst in the first one. Blowing hot air across the catalyst turns the carbon into carbon dioxide (CO_2) and sweeps all the carbon out, a step called *purging*. Meanwhile, the dual reactor is making propylene and having its catalyst slowly incapacitated. Virtually all the propane is converted to propylene. Only about 1 percent or so comes out as by-product methane plus hydrogen.

Isobutane to isobutylene

Petrochemical companies want isobutylene to make MTBE (methyl tertiary butyl ether) and a few other petrochemicals. In some countries, gasoline manufacturers use MTBE as a high-octane gasoline component with the added attraction of an oxygen atom in the middle. That helps achieve complete combustion of gasoline in a car engine, reducing the emissions of unburned hydrocarbons. Again, the trick is to use the right catalyst and operating conditions.

Butane to butadiene

The dehydrogenation process with different catalysts, pressure, and temperature can be used to transform normal butane or butylene into butadiene. Some processes use the calcium nickel phosphate catalyst that has been around since World War II. Newer setups add oxygen to the reaction and use other exotic catalysts to reduce by-products.

The butadiene molecule has the attractive feature of two double bonds between two of the four carbons. Butadiene is used extensively to make rubber products. One of the double bonds in the molecule can be used to make long polybutadiene chains, thousands of butadiene molecules long! Then the other double bond in each former butadiene molecule can be used to connect the long chains of polybutadiene together.

Polybutadiene is a synthetic type of rubber. Inadvertently, Charles Goodyear discovered that sulfur atoms can connect or crosslink the long chains to give a durable rubber. He learned that once he added the sulfur to molten rubber and the molecules crosslinked, the three-dimensional mass was an immutable solid. It was flexible but not moldable. Thereafter, he was careful to crosslink this *thermoset* polymer in a mold of its final shape. Once crosslinked, thermosets cannot be melted and reshaped. That's why they grind up old tires for filler instead of melting them down.

Unfortunately for Goodyear, all he and his descendants got out of it was their name on someone else's company. And on a blimp.

Takeaways from NGL Markets—Petrochemicals[1]

- The NGLs make excellent raw material feed to olefin plants that produce ethylene, propylene, and other petrochemicals.
- In an olefin plant, the heavier the NGL feed, the more by-products are produced.
- The olefin plants built to run naphtha can run the other NGLs, but the plants designed to run ethane cannot run naphtha (natural gasoline) and maybe even propane and butane because of their limited separation section and the larger volumes of by-products.
- PDH converts propane into propylene, with almost no by-products.
- Dehydrogenation is also used to make isobutylene from isobutane and butadiene from butane.

1 If these quick takeaways go in the opposite direction of your curiosity, you can find much more detail in *Petrochemicals in Nontechnical Language*, by Donald L. Burdick and William L. Leffler, Tulsa, OK: PennWell, 2001.

Exercises

1. How many pounds of ethane, propane, butane, or natural gasoline need to be fed to an olefin plant to make a billion pounds of ethylene per year? (Use Table 8–1.)
2. How much is that in barrels per day? (Use the following pounds per gallon: ethane = 2.97; propane = 4.24; butane = 4.87; natural gasoline = 5.4.)
3. The Totally Flexible Chemical Company cracks 100% propane in their olefin plant. The company needs more propylene. What are its alternatives?

CHAPTER 9

NGL Markets—Fuels

If you can't stand the heat, get out of the kitchen.

—Harry S. Truman (1884–1972)

Crystallizing Moments for NGL Fuels

1792 William Murdoch demonstrates gas lighting in his home in Cornwall.

1823 Samuel Brown patents the first internal combustion engine.

1892 John Froelich manufactures the first gasoline-driven, one-cylinder tractor.

1895 Charles Duryea patents a gasoline-powered automobile.

1912 John Gahring installs a propane tank in his home for cooking and heating.

1913 First propane-powered car is built.

1925 The octane of gasoline reaches 60.

1937 Aviation use of butane ends with the Hindenburg zeppelin.

Almost all natural gas liquids (NGLs) get put to the flame, even ethane. Despite the revelations in the previous chapter about the use of ethane as olefin plant feed, that represents only part of ethane's total availability. The ethane content of *all* natural gas streams is hardly known, but the maximum ethane recovery from the best plants is about 90%, as described in chapter 5. Most of the older plants recover half that or less. Many gas plant operators do not try to recover any ethane at all. The unrecovered ethane stays with the methane, and almost all of that ends up at a burner tip in the residential, commercial, or industrial markets.

The use of the recovered ethane has been covered so well in chapter 8, on petrochemicals, that what is left for this chapter are the following:

- The combustion markets for propane, which in the United States consume about half the total production. (The other half goes to olefin plants or exports.)
- Using butanes and natural gasoline to make motor gasoline; only a small amount ends up elsewhere.

Propane

Grades

In the United States there are at least four specifications for propane to choose from. The Gas Processors Association has one for HD-5 propane (GPA Standard 2140.) The 5 in HD-5 refers to the limit on propylene content to 5%. The American Society for Testing Materials uses ASTM Standard B1835. Commercial grades of propane, typically used at petrochemical feedstocks, are either negotiated or comply with the GPA or ASTM standards. The California code of regulations has his own description, called HD 10, as most of its propane comes from refineries with a bit more propane in it.

Residential

The typical home with propane service is in a rural area (fig. 9–1). It is just too far from natural gas lines to warrant a local natural gas distribution company making the investment in underground pipelines to reach it. Sometimes entire rural residential subdivisions (communities) have propane delivered by truck rather than natural gas by pipe.

With natural gas out of the picture, propane competes mostly with electricity for residential customers. That service, of course, starts with coal, oil, or natural gas burned at a remote power plant (or nuclear or a renewable source of energy.) To get to the home, electricity travels down transmission lines and local power grids. Electricity does not have to be "pumped"—it just flows. But line losses take about 30% of the electricity along the way. As improbable as it seems, the flowing electricity heats the lines, losing energy to the environment. In comparison, propane from processing natural gas or from oil refineries may travel by pipeline, transport truck, or bobtail to get to a residence. That involves a different, less investment-intensive cost, but higher operating expenses and its own energy-consumption costs.

Fig. 9–1. Residential propane tank

Unfortunately, when choosing a fuel for a new house, the harried builder or buyer has to grapple with totally obscure unit costs. Propane prices come in dollars per gallon; electricity comes in cents per kilowatt-hour or cents per therm. Then to completely obfuscate the decision process, the efficiencies of burning propane versus using electricity differ by appliance—furnaces, heat pumps, stoves, clothes dryers, and water heaters. Further, the number of really cold days in the winter complicates the comparison of heat pumps and direct-burning furnaces. The efficiency of heat pumps declines as the outside temperature drops. They cannot deliver much heat to a house when the temperature goes down into the 30s (°F). At those temperature there is not much heat to be extracted from the air outside. A house has to have a backup electric resistance system (which may be the most expensive way to heat) or a propane furnace to provide extra heating in cold weather.

Asking a propane dealer if propane is more economical than electricity is like asking a barber if you need a haircut. Ask heating, ventilating, and air conditioning (HVAC) dealers the same question, and they will tell you that electric heat pumps have an efficiency of 240% for heating and can beat propane by a factor of 2.5 (they probably won't add *in nice weather*). In the end, the homebuilder needs detailed engineering calculations, taking into account the geographic location and the prices of propane and electricity, to pick the right fuel system. And those results are good as long as the relative prices of propane and electricity stay the same—which is improbable.

Where there is propane heating, there is most likely a propane stove for cooking, a propane hot water heater, and perhaps a propane clothes dryer.

And where there are any of those, a propane outdoor grill fed by a main tank or a propane cylinder may appear on the patio (fig. 9–2).

Fig. 9–2. Propane outdoor grill

Heat pumps. These devices operate in two modes—cooling in the summer and heating in the winter. When a heat pump operates as an air conditioner, its compressor and evaporator make cold refrigerant. Passing warm indoor air by that cold refrigerant in an inside heat exchanger makes cool air—hence air-conditioning.

In heating mode, a heat pump operates in the opposite direction. The cold refrigerant from the evaporator passes by an outside air heat exchanger where it picks up heat from the environment and then passes it to the house. Heat pumps need a driver, either an electric motor or, in the case of propane, an internal combustion engine connected to the compressor.

Commercial buildings

Heating a commercial facility has all the same issues as residential: efficiencies, geographic location, and prices. Because of the larger scale, propane-fired air-conditioning has some application advantages for commercial and light industrial buildings. Take waterfront hotels in southern coastal areas. They have three issues to be solved: lowering the inside temperature, lowering the inside humidity, and perhaps heating the pool.

Propane-fired cooling seems like an oxymoron, but every air conditioner has a motor or engine to drive a compressor. In this case it is a propane internal combustion engine hooked up to a compressor to drive a heat pump for both summer cooling and winter heating. Most hotels have bathrooms that continually vent air to the outside. That requires incoming air to replace it, with all the humidity it carries, often making lobbies and hallways uncomfortably sticky. Dehumidifiers driven by propane internal combustion engines pass the incoming air through desiccants to pull out the water and then through a chiller as the air is on its way into the halls (fig. 9–3). The air passes under doors, through bathrooms, and out exhaust fans. Desiccants are solid or liquid materials that remove moisture from the air by absorption (in liquid) or by adsorption (on solids.) After running for a while, the desiccant is mechanically "wrung out," then put back in service. The waste heat from the engine can be used to heat pool water.

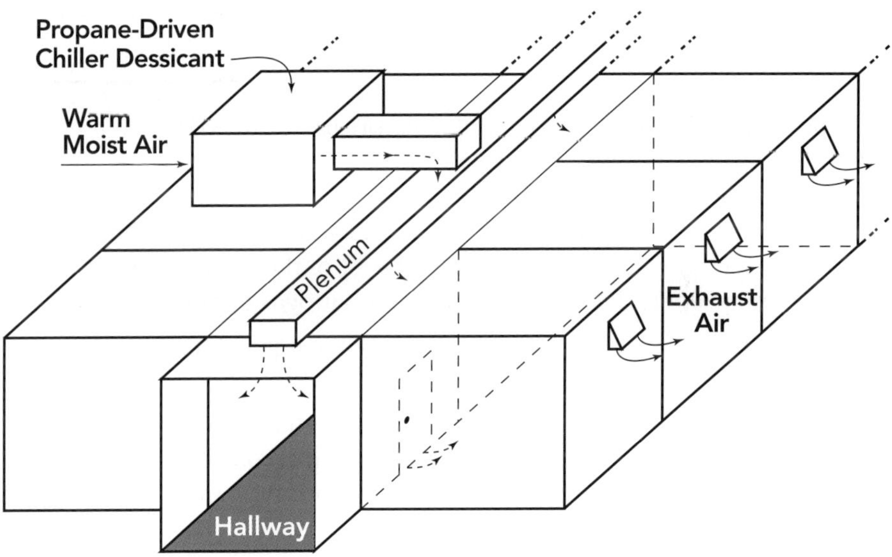

Fig. 9–3. Propane-fired dehumidifier

Generators and combined heat and power

A handful of companies manufacture propane-fired generators, which are internal combustion engines that drive a device that converts mechanical power to electricity that flows through a circuit external to the applications.

Once available only to industrial applications, combined heat and power units (fig. 9–4), sometimes called CHP units, now come in residential and commercial building sizes. CHP units capture exhaust from the propane engine to raise the temperature of the water that circulates, in a way that heats a room. Other fluids besides water can also be used to for this purpose.

CHP units may have regulators that can operate at 120 or 240 volts and at a 60-Hz output regardless voltage output. Smaller ones are air-cooled; larger ones have a water-coolant radiator.

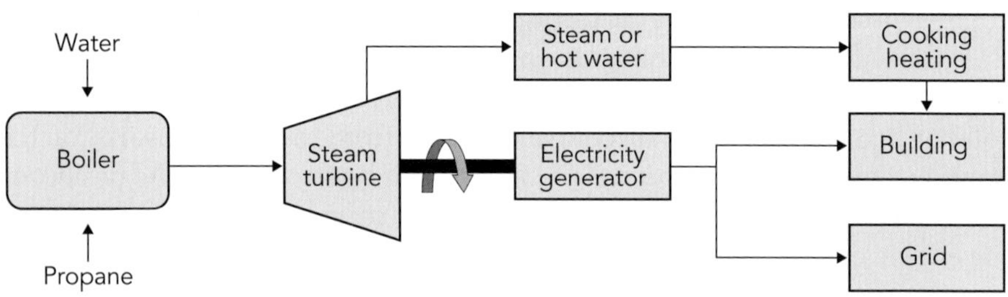

Fig. 9–4. Diagram of a CHP unit

Automotive fuel

Some countries (e.g., Turkey and Poland) use liquefied petroleum gas (LPG) (sometimes propane, sometimes propane/butane mixes) more than any other fuel for their cars. In other countries (e.g., the United States), propane is only a small part of the mix. Worldwide, LPG is the third largest automotive fuel by volume, but still only 2%–3% of the total across the world. Drivers in most countries usually refer to it as autogas or autoLPG.

Propane has a few advantages over the other liquid fuels, gasoline and diesel fuel:

- Emissions of CO_2 are 35% less per gallon than for gasoline and even better than for diesel fuel. That calculation is complicated by the fact that propane yields fewer miles per gallon. Propane has only about 73% of the energy content of gasoline.
- Propane burns cleaner than other liquid fuels. Because of its simpler molecular structure, there are few of the trace elements that are found in gasoline and diesel that cause gunk accumulation on fuel injectors and valves.
- In many countries, governments use fuel taxes to accomplish social engineering. They lower taxes on autogas and other fuels. Combined with local market conditions, that makes the cost per mile of autogas an attractive alternative, especially in regions where natural gas prices are relatively much lower than oil prices.

- Propane has an octane rating of 107, which makes it suitable for powerful high-compression engines.
- The low boiling point of propane, −43.8°F, gives propane an advantage in cold weather. In order for an internal combustion engine to start up in frigid conditions, some vapor has to be injected into the cylinder. For gasoline engines, butane will have been added to the fuel mix. For diesel engines, chemical additives help atomize the fuel injection. Propane has neither problem since it easily vaporizes at any humanly tolerable low temperature. Butane has a boiling point of 31.1°F, which might suggest problems in frigid weather, but it benefits from a symbiosis with propane in an LPG of propane/butane mix.

Not every aspect of propane as an automobile fuel is favorable, though:

- In many countries, including the United States, refueling stations are sparse, making long-distance travel problematic and refueling for casual drivers inconvenient. Using propane in commercial vehicles that return to a single location every day allows for easier refueling.
- Fuel tanks in cars generally sit in the trunk (called the "boot" in some countries) and take up storage space. Many car models are designed to run on both autogas and gasoline, switching at the turn of a dial. That, of course, calls for two fuel tanks.
- Because of the high pressure to keep propane liquid, the fuel tank requires heavy-duty construction and more weight. Also, to achieve the same range as the other fuels, the tank needs to be larger.
- Compared to other fuels, service personnel qualified to deal with propane vehicles are fewer in number, and so servicing is less convenient.
- Autogas involved in collisions is clearly a more dangerous fuel.

Manufacturers. Scores of companies make original-equipment autogas-fueled cars: Citroen, Fiat, Ford, Hyundai, General Motors, Suzuki, Peugeot, Renault, Sköda, Tata, Toyota, Volvo, and Volkswagen. Most of them also make the dual-fueled cars that run on autogas and gasoline.

Vehicles. Autogas cars have the same hardware as other internal combustion vehicles, except for the equipment for mixing the fuel with air before it enters the cylinders. Many autogas cars still use technology that has been around since the 1940s: the liquid autogas in the line from the tank goes to a converter-mixer that releases an ignitable autogas/air mixture to the engine's intake valves and then into the cylinder.

A newer system mimics fuel-injection technology. Autogas goes to a converter where liquid goes to vapor. The gaseous autogas then goes into the air intake manifold under computer-controlled conditions. The autogas-spiked air then goes to the engine intake valve and into the cylinders.

Refueling. At refueling stations the fuel tank has to be hooked up to the supply in a totally closed system (fig. 9–5). Different countries have their own nozzle fittings. In some places, the pump nozzle has to be screwed onto a threaded fitting on the car's fill line. In other places, a bayonet-type fitting attaches by a push and twist. Both have check valves that shut off upon disconnecting so that no autogas can escape to the atmosphere

Fittings on the car and the pump nozzle are typically made of brass because steel can spark. That is not something anybody wants around autogas.

Fig. 9–5. Propane dispenser

Applications. Centralized refueling makes autogas appropriate for many kinds of fleets:

- Police cars and ambulances that "return to the barn" at the end of the shift are good candidates for autogas. Politicians and bureaucrats see autogas fueling for these vehicles as role-modeling for an improved environment.

- School buses are run by school districts that perennially never have enough funding. That makes the fuel and maintenance cost savings of autogas attractive. In addition, autogas being a greener fuel impresses their young audience.
- Lawnmowers for large expanses such as parks and other green spaces may cost a little more than the alternative. The manufacturers have not yet reached the same economies of scale. But the cheaper fuel quickly offsets the front-end investment.

Forklifts

Forklifts that run on autogas have decided advantages in warehouses and factories. The competing fuels are gasoline and electricity. Gasoline engines on forklifts emit much more toxic vapors than those using autogas. That mitigates a crucial human resources issue, the quality of the workplace environment.

Of course, electric motor–driven forklifts work fine, but in many cases they are in service for hours extending beyond one shift, which is the limit of a charged forklift battery. That calls for either a 2-hour battery recharge or changing out bulky battery packs, a labor-intensive effort. Autogas-powered forklifts can be refueled by hose in about 10 minutes or by detaching and removing the propane cylinder (fig. 9–6) and replacing it. They generally move faster than electric forklifts, and they do not run down as the battery power gets used up.

Fig. 9–6. Propane-fueled forklift

Weed control

Farmers and railroads have long depended on herbicides control weed growth. Close scrutiny of the long-term effects on the environment and in humans of these chemicals has led many to turn to nontoxic alternatives, especially given the rising popularity of organic crops. Tractor-drawn flame generators carefully apply heat to the weeds, not burning them, but increasing the internal temperature of the plant cells (fig. 9–7). That causes them to rupture. The loss of water from the cells and a change in protein structure does the weed in and it withers and dies. Corn, soybeans, sunflowers, orchards, and blueberries crops are amenable to propane weed control.

Fig. 9–7. Flame weed control in a sunflower patch

Besides its use with edible crops, propane weed control is good for other toxic-free spaces such as parks and playgrounds. Handheld and pushcart sizes are used in smaller scale applications. Also, railcars equipped with burners keep the railway ballast free of weed growth.

Crop drying

Propane provides fuel for crop drying that takes place pleasantly enough in the off-season when home and commercial heating loads are down. Crops that need drying include the cereals (wheat, oats, barley), corn, rye, soybeans, tobacco, and even potatoes. The cereals come off the field with as much is 15% too much moisture. A propane-fired air dryer reduces the excess moisture, reducing spoilage due to mold and mildew, and kills most of the resident critters (fig. 9–8). All this allows for longer storage and higher-quality product and prices. Propane burns cleanly enough, compared with other liquid fuels, that it can be combusted directly in the air going to a grain dryer. That makes propane heating 100% efficient. As for potatoes, they sweat after harvest and need warm air to remove the accumulated moisture.

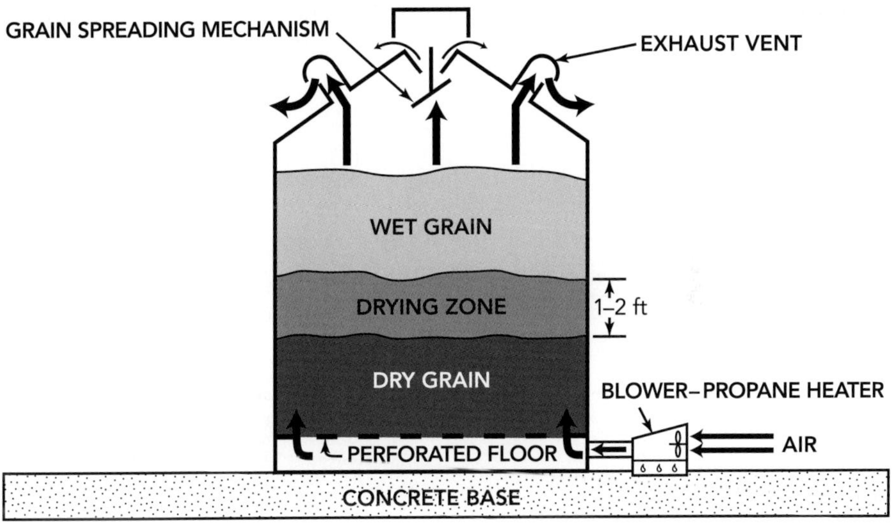

Fig. 9–8. Grain drying

Curing tobacco needs hot air circulating among the leaves. Propane used to be burned right in the curing barns until it was found that the carbon dioxide in the combustion gas mixed with some tobacco components and created nitrosamines, a known carcinogen. Now air is heated outside the barns in a heat exchanger and pumped in. That is one less worry for tobacco users and the companies that sell tobacco products.

Other agricultural uses

Farms are huge generators of toxic waste products—chicken guts and feathers, dead or sick (and soon to be dead) animals, excrement, contaminated crops, and other disgustingly undesirable matter. Propane incineration provides a sanitary way to dispose of it all. Further addressing particularly revolting actions, propane flame burners also kill pests and burn waste in poultry houses and kill flies and other insects in the muck in stockyards. There is no residue to pollute the soil, and no special licenses or protective gear is needed. No nearby residents have to be notified, and no field entry bans have to be posted. (Why don't all farmers use flame burning?)

During the winter months, farmers closely calculate the trade-offs among livestock weight levels, feed intake, and barn temperatures. Propane heating provides an affordable way to deal with these decisions.

Irrigation pumps and electricity generators abound in agriculture, and propane-driven combustion engines drive both. Propane has an endurance advantage over gasoline and diesel for these applications. Since propane contains virtually no olefins, as gasoline and diesel do, it can be stored indefinitely at the customers' locations. Gasoline and diesel will begin to form gums and varnishes after a few months in a storage vessel.

Balloons

The list of applications would hardly seem complete without the addition of (so-called) hot air balloons. Each of the balloons at the annual Albuquerque International Balloon Fiesta (fig. 9–9) has a propane tank and burner. It provides the hot combustion gases, CO_2 and water vapor, that are lighter than air. Filling the balloon with these (a little hot air comes along) lifts the playful vehicle into the sky.

Fig. 9–9. Propane hot air balloons at the Albuquerque International Balloon Fiesta

Butanes and Natural Gasoline

Any owner of a gasoline-powered car who lives in Bemidji, Minnesota, has a challenge every winter—starting the car in subzero temperatures. The fact is, liquids do not burn. Their vapors burn. Getting the reaction between a hydrocarbon and air requires an intimate mixture of the fuel molecules and the oxygen content. That is why gasoline is a dangerous fuel—it has vapors coming off all the time. Diesel fuel has hardly any. Gasoline has a low flashpoint; diesel has a high flashpoint.

Now back to Bemidji. In order to get a car started on a February morning, some vapors have to be injected in the engine cylinder, along with the rest of the frigid liquid gasoline droplets plus the air in order to get a combustible mixture for the spark to ignite. Butane is a component that allows that to happen. As the gasoline/air mixture gets injected, butane provides the necessary vapor. As the butane ignites when the spark plug flashes, the rest of the gasoline will vaporize and then burn as well.

All this is taken into account when refineries blend gasoline for cold-weather use. Governmental organizations such as the U.S. Environmental Protection Agency (EPA) limit the amount of butane that refineries can put in gasoline, in order to mitigate any spurious leaks or other emissions of butane to the atmosphere at gas stations or from cars' leaky seals or other engine problems. Regulations set the volatility of the gasoline measured by the Reid vapor pressure (RVP). All the gasoline-blending components generated by a refinery have unique RVPs. Generally, butane is the least expensive and is also the most volatile, at 52.0 RVP. Refiners have an incentive to put in as much butane as the RVP limit allows.

A gasoline-blending example

Table 9–1 shows an example of how refiners calculate the amount of normal butane to add to a blend of other components from around the refinery.

Lines 1 through 5 are the given gasoline components. Line 6 shows the calculation of the weighted average RVP, using the last column, the sum of the product of the volumes times their RVPs, divided by the total volume.

To calculate the RVP of the five components, divide the weighted average by the total volume: 59,600 ÷ 11,000 = 5.4 RVP (rounded).

The next steps calculate how much butane has to be added to the blend of the five components to bring the RVP up to 9 RVP in the example. Line 7 is the blend of the five components again. Line 8 has the amount of butane, X, and the butane RVP of 52.0. Using that same algebraic technique gives X = 4,990, the amount of butane that brings the blend up to 9 RVP.

Table 9–1. Blending gasoline to meet RVP specifications

		Volume (barrels)	RVP	Volume × RVP
1	Cat cracked gasoline	5,000	4.4	22,000
2	Reformate	2,000	2.8	5,600
3	Alkylate	2,500	4.0	10,000
4	Natural gasoline	500	6.0	3,000
5	Ethanol	1,000	18.0	18,000
6	Blend totals	11,000	?	59,600

To get the RVP of the blend of the five components

59,600 ÷ 11,000 = 5.4 RVP (rounded)

To get how much normal butane, call it X, is needed to get the RVP up to, say, 9 RVP:

7	Blend total	59,600	5.4	321,840
8	Normal butane	X	52	52X
9		59,600 + X	9.0	321,840 + 52X

So, solving for 9.0 × (59,600 + X) = 321,840 + 52X,

X = 4,990 barrels

By the way, if the automobile owner lives in Del Rio, Texas, driving a car during the dog days of August presents a different problem. Temperatures can climb into the triple digits. With too much volatility, gasoline could vaporize in the liquid flow line from the tank to the engine, starving the fuel injector and shutting down the engine.

Because cars do not need much butane to achieve combustion in Del Rio, and because the high temperatures exacerbate the emissions problem, the EPA limits the RVP in that region to much lower levels than Bemidji. The EPA sets limits by geographic area as well as by season.

Normal butane has the highest RVP of all the petroleum components and is abundant and cheap. (Ethanol has a higher RVP and is more expensive.) That leads refiners to think of normal butane as the "swing" pressurizing agent, the one to bring the RVP up to its maximum allowable level.

Isobutane

In most refineries, most or all of the isobutane goes to an alkylation plant, where it reacts (and loses its molecular identity) with butylene and propylene to make alkylate, a high-octane, gasoline-blending component.

Isobutane is also used in a number of smaller volume but somewhat exotic-sounding applications. For example, people lighting up with their Bic lighters are burning isobutane. Butane lighters use isobutane instead of normal butane because the boiling point (temperature at which vapors start to come off) is lower, 10.9°F versus 31.1°F, and the volatility (vapor pressure) is higher, 71 RVP versus 53 RVP.

Octane number

So what are octane numbers anyway? Everybody picks one whenever they fill up their car, but few people really know what they are getting. Understanding octane numbers needs understanding of how an internal combustion engine works, so here goes.

After a gasoline/air vapor mixture is injected into an engine's cylinder, it is compressed by the piston. At the top of the piston stroke, the spark plug lights off, causing the gasoline/air mixture to ignite. The gasoline burns almost explosively and causes the combustion gases, CO_2 and water, to expand. That drives the piston down the cylinder in the power stroke. A connecting rod transmits the power to the power train.

Compression causes temperatures to rapidly increase. (Pumping up a bicycle tire does the same thing. The bottom of the pump feels warm at the end of the exercise.) As the engine's piston compresses the gasoline and air, the vapor could get so hot that it could self-ignite before the piston completes its upward thrust. If that happens, the burning, expanding gases push down on the piston as the engine tries to push the piston up. The driver might hear pinging or knocking, which is bad for the engine because it rattles the parts and it reduces the miles per gallon as the engine works against itself.

A test engine determines the octane number of a gasoline or gasoline-blending component by measuring the compression ratio, how much the gasoline/air mixture can be compressed, before knocking occurs. Big cars with big compression ratios need higher octane numbers than smaller cars need.

Perhaps it would be a surprise to most people that they probably use isobutane every day. When the chlorofluorocarbon (CFC) ban started in the 1990s (Freon was the most notorious villain because of its ozone-destruction tendencies), isobutane became the propellant of choice in many brands of hairspray, cooking spray, and shaving cream. It is also used in many refrigeration systems as the gas that expands and cools, and later compresses and then rejects heat. Isobutane is also used for calorimetric measurements, calibration of gas mixtures, and emissions monitoring.

Octane numbers

As refiners put butane and natural gasoline into motor gasoline blends, they have to take into account octane numbers. As before, every gasoline-blending component has a unique octane number.

The weighted average of the octane numbers of the components gives the octane number of the blend.[1] Normal butane has an octane number that makes it easy to fit into motor gasoline. Natural gasoline has a particularly low number. Refiners often work it over chemically to increase its octane number: Some refiners send the pentane and hexane parts of natural gasoline to the C_5/C_6 isomerization unit (covered in detail in the refinery chapter). They do this to improve the octane number so they do not have to add as much of another high-octane component like alkylate to offset natural gasoline's poor octane rating as they blend it into 87, 89, or 91 octane gasoline.

Natural gasoline as diluent

Natural gasoline is sometimes used as a diluent to facilitate the movement of very heavy oil. The bitumen at some Canadian operations is so viscous that it cannot be pumped through a pipeline. The producers add natural gasoline to bring down the viscosity, making the bitumen pumpable. It's almost like putting hot water in dry oatmeal to make it swallowable. When that diluted bitumen gets to a refinery, the natural gasoline comes out of the distilling unit with the straight-run gasoline and is indistinguishable from it.

Takeaways from NGL Markets—Fuels

- The ubiquitous fuel, propane, finds markets in residential, commercial, agricultural, automotive, and recreational uses. Almost all of them are in exurban markets where the alternative, natural gas, is not available.
- Most normal butane usually ends up as a motor gasoline component, added in amounts to bring the vapor pressure of the gasoline blend up to the maximum allowed by regulation.
- Almost all isobutane is used in a refinery alkylation plant to make high-octane gasoline blending components.
- Natural gasoline also ends up in motor gasoline after it is treated in a refinery. It has a low octane number, and refiners may process it further in an isomerization unit or compensate by adding other higher octane components to the blend to meet market specifications.

---

1 Actually, the octane numbers of individual gasoline-blending components don't always blend that way when mixed with other components. That's why refiners developed "blending octane numbers" from experience to better reflect how the components behave in a blend. The numbers are close, but not exactly the same.

Exercises

1. How much normal butane has to be added to 10,000 barrels of natural gasoline to bring the vapor pressure up to 9.0 RVP? (Use the data from table 9–1.)
2. How about the following blend?
   - 1,000 barrels of cat cracked gasoline
   - 2,000 barrels of reformate
   - 1,000 barrels of alkylate
   - 400 barrels of ethanol (RVP = 18)
3. Name at least three advantages and three disadvantages of fueling a car with propane.

CHAPTER 10

Safety, Operations, and Maintenance

A great flame follows a little spark.

—Dante, *The Divine Comedy, Paradiso*

On July 4, 1967, in Ozona, Texas, John Goodall and Harley Nesbitt were riding in a pickup truck on their way home from their jobs at the local gas plant. Beneath the road they traveled, a butane line had corroded and leaked enough gas to create a large, ground-level butane cloud. As Goodall drove into the cloud, the truck stalled when butane entered the pickup's manifold, starving the engine of oxygen. Unaware of the hazard, he cranked the engine to restart it, and a spark, probably from the starter, ignited the butane. The explosion and flames instantly consumed the truck, its occupants, and vegetation within a 50-yard radius.

Firefighters arrived on the scorched scene in 10 minutes, but could not ascertain the source of the fire for a half hour. Another half hour passed before the butane line was shut down. The fire dissipated after all the butane in the blocked line vented.

Safety

This is not the kind of book that should be considered a guide for safe handling of natural gas liquids (NGLs). Not that liability issues are inconsequential, but this book is meant for people who do not handle NGLs but who deal with people who do. That's why *you* bought this book, right?

Still, safe handling of NGLs takes center stage at every point along the value chain, from production to the burner tip, from rig operator to homeowner. To demonstrate the kind of hazards and risks every operator may encounter along the way, the following sections describe some circumstances that just did not work out favorably for the operators involved. The scenarios borrow heavily from information published by the Propane Education and Research Council (PERC), which provides extensive help to the industry on safety, marketing, market development, and other aspects of the business. The intent here is to demonstrate why responsible companies have continual safety training for every operator. As the gurus say, "Safety is mission critical."

Most of the examples involve propane. As mobile and ubiquitous as that NGL is, people and circumstances will find many ways to cause hazardous incidents such as the following.

BBQ on fire!

The propane supply under a BBQ grill has sprung a leak and the propane has ignited, threatening the adjacent house. The heat from the fire has raised the temperature and pressure of the contained propane, causing the pressure relief valve on the cylinder to open, adding more fuel to the fire.

BLEVE

Propane and butane moving by tank car or tank truck, or even in pipelines, present a particularly dangerous hazard if the container is punctured: a BLEVE (boiling liquid expanding vapor explosion). If one ruptures, the sudden release of pressure allows the liquid propane or butane to instantaneously begin rapid vaporization. If the expanding vapor comes into contact with an ignition source—a spark from the puncture, a car engine intake, a cigarette, or even static electricity—as it inevitably does, the resulting fire will cause the vaporization to compound, escalating to explosion and conflagration (fig. 10–1).

Firefighters responded and immediately directed a water stream at the propane cylinder to cool the contents, in order to let the relief valve reset itself in the closed position. Then, using flame protection and an extension tool, they closed the valve on the cylinder to cut off the remaining fuel source. They had to use a pipe wrench to turn the valve stem since the handle was damaged.

Fig. 10–1. A propane boiling liquid expanding vapor explosion (BLEVE) at the point of explosion (courtesy of the Booneville Fire Department)

Rear-ended bobtail delivery truck

As a bobtail truck stops at a red light, the 18-wheeler behind it slides into it. The fittings on the back of the bobtail are compromised. Leaking propane immediately ignites. The driver of the 18-wheeler is slumped over the steering wheel, but the bobtail driver is unhurt.

The bobtail driver immediately activates the truck's emergency shutoff valve located away from the rear end of the truck. The various lines at the rear of the bobtail that contain about 10 gallons of liquid propane continue to evacuate into the ongoing flame, but that amount of propane burns itself out in about 15 minutes. Meanwhile, firefighters and emergency medical technicians (EMTs) attack on two fronts. EMTs attend to the driver of the 18-wheeler; firefighters check with the bobtail driver to ensure that the emergency shutoff valve is closed and ask if the driver knows what might be the source of the leak. Meanwhile, they direct water on the flame source and on the shell of the bobtail to keep it cool and prevent the pressure relief valve from popping.

Overturned propane railcar on fire

Several railcars have left the tracks and turned on their sides. An acetone car is leaking badly and is feeding a fire under a propane car. The jumbo propane car, with about 32,000 gallons, lies in a way that the pressure relief valve at the top is below the liquid level line. It is releasing liquid propane, which adds more fuel to the fire. The incident site is about 500 feet from a county road.

Due to the remote location of the accident, the burning propane tank car has a reasonable probability of rupturing into a BLEVE. This is exacerbated by the lack of any cooling effect that vapor release from the pressure relief valve might have had. The local authorities immediately evacuate the area within a 3,000-foot radius and notify the railroad to stop all traffic on that line. If sufficient water supply had been available, a steady stream on the propane tank car might have reduced the temperature enough to close the pressure relief valve, but the extent of the fire does not warrant putting firefighters in harm's way. After an hour, the propane tank car does rupture in a dramatically roiling fireball and then burns itself out in another three hours.

Leaking propane-fueled car

A security officer at a mall smells a strong mercaptan odor in the underground parking garage and immediately calls 911 for fire department assistance. On arrival, the fire chief sends in two firefighters with a handheld device that detects combustible gas. They find dangerous levels of gas in the garage and eventually home in on high levels of explosive vapor around a car with a diamond-shaped sticker on the back with the word "propane" on it (fig. 10–2). The area is evacuated. Firefighters bring in blowers to dissipate the vapors.

Fig. 10–2. Propane car sticker

The National Fire Protection Association (NFPA) code calls for propane-fueled cars to have a "propane" label on the back. These type cars have various fittings that handle and have access to the fuel, including a fill line with liquid and vapor lines, a pressure relief line, a tank gauge, and a fixed maximum liquid level gauge. Any one of these could be a problem. In addition, the problem could be coming from a loose propane cylinder stored in the car that has nothing to do with the fuel system.

The source with highest probability of being the propane leak is in the trunk. The firefighters first go to the engine compartment and disconnect the battery to reduce the likelihood of a spark when opening the trunk. In the trunk they remove the valve housing cover and look for frost, a sign that propane is leaking, causing low temperatures and ambient humidity to freeze. They use soap and water to detect and pinpoint leaks. If the trunk is not the source, they shut off the main supply valve and look in the engine compartment, using the same procedure. Even though the main supply valve is shut off, the system downstream of that valve still has liquid or gaseous propane in it and it will take a while to pressure down. The car is eventually towed to a safe environment for repair.

Underground tank compromised by a lawn service

While mowing his the lawn, a new gardener drives a power mower over some fittings in the heavy grass. Vapors with a strong odor spew from the fittings. The gardener jumps from the mower as the vapors ignite. A large fountain of fire envelops the mower, giving off black smoke as the tires burn. The gardener calls the fire department.

The riser from an underground propane tank has fittings at the top, at surface level, to operate and service the tank. That includes liquid fill and vapor lines, a pressure relief valve, a fixed maximum liquid level gauge, a service valve, and maybe a pressure gauge. The riser sits inside a protective shroud (fig. 10–3).

On arrival, the firefighters ascertain the source of the fire is a riser from an underground propane tank that is located about 2 feet below the surface. They call the dealer for technical information about the tank. They begin hosing down the house to protect it, and they clear the area of the accumulated spectators. From appearances, the riser and a line have cracked somewhere.

Given the inability to approach the tank because of the flames, the firefighters opt to let the propane fire burn until the fuel source exhausts itself. Initially, rapid vaporization causes a large blaze. In the process, as the propane turns from liquid to vapor, the cooling effect lowers the temperature of the liquid in the tank, slowing down the rate of vaporization. Eventually, the leak reaches a steady state of low volume and low flame. The ground around the tank freezes, and with the vaporization keeping the propane cold, there is little chance of the underground tank rupturing, causing a BLEVE. When the flame eventually burns out, repair personnel in protective clothing, accompanied by a protective sprayed water mist, excavate and determine what repairs are necessary to fix the tank.

Fig. 10–3. Typical residential underground tank. The parcels underneath provide grounding and corrosion control (courtesy of Blossman Propane).

Forklift propane tank fittings fail

At a warehouse loading dock, a forklift operator notices the smell of smoke as she backs up. On inspection, she sees that loose battery cables have come into contact with the engine's intake manifold where propane mixes with air. An electric spark started a fire. She shuts down the forklift and goes to get a fire extinguisher. Another nearby employee attempts to remove the propane fuel cylinder by using the quick disconnect on the fuel line, but only manages to open the service valve on the cylinder to the atmosphere. The propane vapor ignites in a flashback, burning the employee. The forklift operator empties the fire extinguisher on the flames, to no avail, and then pulls the burned employee away from the forklift. Meanwhile, someone has called 911 for emergency services.

Propane cylinders mounted on the back of forklifts give as many as five accesses to the propane at their business end, including a pressure relief valve and the service valves for filling and delivery. The service or delivery valve sends liquid propane to the manifold in the forklift engine compartment.

On arrival, the firefighters evacuate the area, move the injured person to a safe area, and put a water hose on the cylinder to keep it cool to prevent rupture. They then let the propane burn until it totally evacuates the cylinder. Meanwhile, water hoses contain the fire and prevent it from spreading to combustibles in the area.

Propane tanker truck overturns

While driving down a highway, a tanker truck driver has to swerve to miss an erratic automobile driver. The truck jackknifes, turns on its side, and slides along the highway. When the emergency response arrives, they see a hazardous material placard on the back of the trailer (fig. 10–4). The driver confirms that the load is propane and that the tank is 80% full. They see no breach of the propane tank, but see a small leak from the bottom liquid unloading valve.

Fig. 10–4. Flammable liquid placard

Propane tankers are made from steel sheets welded into a cylinder. The ends have stamped spherical or elliptical heads. They have pressure relief valves; remote liquid and vapor transfer lines; temperature, pressure, and liquid level gauges; and emergency valve controls. The transfer lines are located on the bottom; the pressure relief valve is located at the top.

After assessing the integrity of the cargo tank, the first effort is to stop the leak at the loading line. The second is to keep the truck in a safe condition until the propane can be offloaded or the truck can be turned upright. The hazardous materials response team is called in to assist. While the firefighters hose down the leak with water to disperse the vapors, the hazmat team members wearing protective clothing try to close the leaky valve. Since it does not close, they try a fiberglass wrap for a gas leak or a freeze wrap for a liquid leak. The ambient temperature is above 90°F, so the team hoses down the cargo tank to keep the temperature from rising to the point where the pressure relief valves pops off. Eventually, another transport truck arrives to offload the propane cargo so the truck can be set upright.

Farm tank flow line breaks

A propane delivery line has an ignited leak at the bottom of a stand-alone tank. The flames have begun to heat up the contents, the pressure relief valve opens intermittently, and the vapors ignite as well.

The standard propane supply tank has all the fittings at the top in a compartment. The most vulnerable fitting is the delivery line that drops over the side to the ground and moves underground to wherever the fuel application is.

The emergency response team begins spraying the tank to keep the contents cool and stop the pressure relief valve from opening. Operating at a farm location may present a water supply issue. A team member must approach the tank and turn off the service valve to stop the flow of propane and thus extinguish the fire. To do that, water hoses, other than the one directed at the tank, provide the firefighters with special cover, called a fog pattern or power cone. Once the valve is closed by the designated team member, the fire burns out shortly. The procedure demands an adequate supply of water and hoses for cooling and protection, including backup if the main supply fails.

Careless homeowner blunders

Homeowners are the number one propane hazard. In a typical example, a homeowner neglects to fully turn off the valve on his propane grill and compounds the problem by not tightening the valve on the propane cylinder, and then almost causes disaster by wheeling the grill into his garage. Later on, after smelling the telltale odor of mercaptan, he calls the fire department and wisely evacuates his house.

The fire responders quickly determine from him the source of the propane vapors, and they open house windows and doors and the garage access door to start ventilation. They use fans to blow the air through the interior. The fans are situated to provide positive pressure, not suction, to prevent any possible ignition source from the fan motor. Using a handheld device that measures the presence of combustible gas, they wait until the propane gas level is low enough that they can enter the garage, turn off the two valves, and move the grill to the outdoors.

Operations

Every point along the NGL value chain, from a gas plant or a refinery to the burner tip, requires both common and unique operating procedures. Locations with massive bulk storage, gas plants, refineries, pipeline terminals, and marine terminals have the usual transfers in and out of tanks. In addition, they continuously load or unload tank cars, tank trucks, pipelines, or seagoing vessels. Transferring to and from these mobile units presents some of the riskiest industrial NGL operations.

Bulk plants

Every bulk plant has a readily available operations manual that details all sorts of procedures involving transferring NGLs between containers. The prose here, of course, does not replace it.

Loading a transport truck. The preliminary steps ensure that all the valves and gauges are operating correctly and that the vehicle wheels have chocks to keep the car immobile. After hooking up the hoses, the operator may do a sniff test to make sure that stenched (odorized) propane is being loaded. (The operator vents a little propane to the atmosphere and smells it.) Nobody smokes in the area.

Pushing propane or butane into a truck can use either a liquid pump or a gas compressor. With a pump, the liquid is pumped via the fill line and excess vapors in the truck go to the bulk storage via a vapor equalizer line between the truck and the bulk tank. When the plant uses a compressor to load a truck or tank car, vapors drawn from the truck's tank are compressed and sent to the bulk plant's tank. As the pressure in the bulk plant's tank increases, it forces liquid propane out through the fill line into the truck's tank.

After the transfer finishes, the hoses are bled down so that they contain vapor only. All hoses have shutoff devices to keep the vapors contained when they are disconnected from the truck's tanks in the bulk plant's tanks.

Unloading trucks at bulk plants. All the essential preliminary steps apply including inspections of the equipment, gauges, tank inventories, and so on. Moving propane and butane into bulk storage is generally by pumping liquid. The vapor equalizer line between the bulk plant's tank and the truck's tank is still required. As always the operator monitors liquid level gauges at both ends during the transfer.

Loading and unloading railcars. The operator secures the area around the rail spur by placing a sign that says "Stop—tank car connected" so that nothing comes down the track to bump the loading/unloading car.

Loading/unloading fittings are in the top dome of a railcar. The operator accesses them by a platform next to the tracks, lowered in place against the car. The operator opens a dome and attaches the safety shutoff valves, liquid and vapor flow lines, and then handles the exchange of propane or butane in the same manner as a tank truck.

Handling cylinders. No matter the size of propane or butane cylinder, it needs inspection every few years. Every cylinder should have a requalification date indicating the due date for one of three inspections:

- Volumetric expansion test—every 12 years
- Proof pressure test—every 7 years
- External visual inspection—every 5 years

In the United States a federally mandated requalification identification number (RIN) appears on the cylinder with the type of inspection and the due date. The inspectors examine the usual list of mechanical features, including gauges, valves, protective structure, and evidence of leaks.

A truck or railcar loaded with propane or butane in bulk or in cylinders has to have a placard (fig. 10–4) clearly displayed that identifies the fact that a hazardous material is moving down the highway. It is no coincidence they are colored bright red.

Residential and commercial operations. To the homeowner and businessperson, propane coming into the home or building and burning in a furnace or on a stove is uneventful . . . until a strong skunklike odor permeates the area. When that happens, some people, but not many, know the urgent procedure:

- Immediately instruct everyone to vacate the premises.
- Do not turn on or off anything that has a switch—lights, telephones, computers, even flashlights.
- Use a phone or cell phone outside the building to call 911 and the propane supplier.

Many propane suppliers also give this advice:

- If the flame on a stove or at a furnace is consistently yellow, not blue, a dangerous combustion problem exists with either the propane content or the appliance. Call the dealer.
- Always have a fire extinguisher nearby.

- Don't have propane tanks or cylinders near sources of ignition such as air-conditioners, fireplace, a fire pit, or an automobile. Leaks happen.

Vehicles. Propane vehicles, trucks, cars, and especially recreational vehicles bump along highways, shaking their tanks, fittings, and valves. Since they operate mostly in the open air, the leaks sometimes go undetected. The risk compounds when they are parked in garages or poorly ventilated enclosures. That calls for servicing by qualified mechanics more frequently than conventionally fueled vehicles.

Static electricity

On cold, dry afternoons, walking across a wool carpet to open the front door can be shocking as a spark jumps between the hand and the door knob. That release of electricity comes from a buildup of static electricity in the human body that can reach an ungraspable 35,000 volts. The discharge of that voltage measures but 35 millijoules, a diminutive-sounding amount. But it is well above the threshold for igniting propane or even natural gasoline vapors. Tank trucks loading propane can accumulate enough static electricity to emit 2,250 millijoules.

Both the American Petroleum Institute (API) and the NFPA have publications that deal with the protocols for eliminating electrical discharges from built-up static electricity:

- NFPA 77: Recommended Practices on Static Electricity
- API RP 2003: Protection Against Ignition Arising Out of Static Electricity, Lightning, and Stray Currents

These publications cover loading and unloading tank trucks and railcars and filling bottles and tanks. They deal with methods to avoid the buildup of voltage. In the end, the most cost-effective way of dealing with the hazardous is through grounding and bonding.

Grounding. Where is the safest place to send all the built-up voltage that comes from flowing fuel, dragging hoses, trucks moving down the road, or railcars down the line? The Earth. It has an inexhaustible ability to receive electrical charges. That's why it's called "grounding": The voltage to be discharged is connected to an electrode stuck in the ground. The connections are critical. The lines are usually copper or steel, but even those have some resistance. Lines that are too long can have enough resistance to prevent the static electricity from totally dissipating. Both the NFPA and API recommend that the resistance of these lines measure no more than 10 ohms end to end.

Bonding. Railcars and tank cars have many parts, metal and otherwise. To discharge the static electricity by grounding, all the parts have to be connected—bonded—in a way that electricity can flow from each to the ground. That prevents one part from discharging to another even though they do not touch, just like a charged hand on a doorknob receives a startling result. That spark happened in nanoseconds before contact between the two as the charge jumped across air.

Tank trucks. Loading procedure during transfers calls for the driver to connect the truck to Earth (grounding) before any other operations. The grounding system should have circuits that prevent transfer of the fuel if the ground connection has not been made. Typically, the loading rack itself has a grounding system that the truck can be connected to (fig. 10–5).

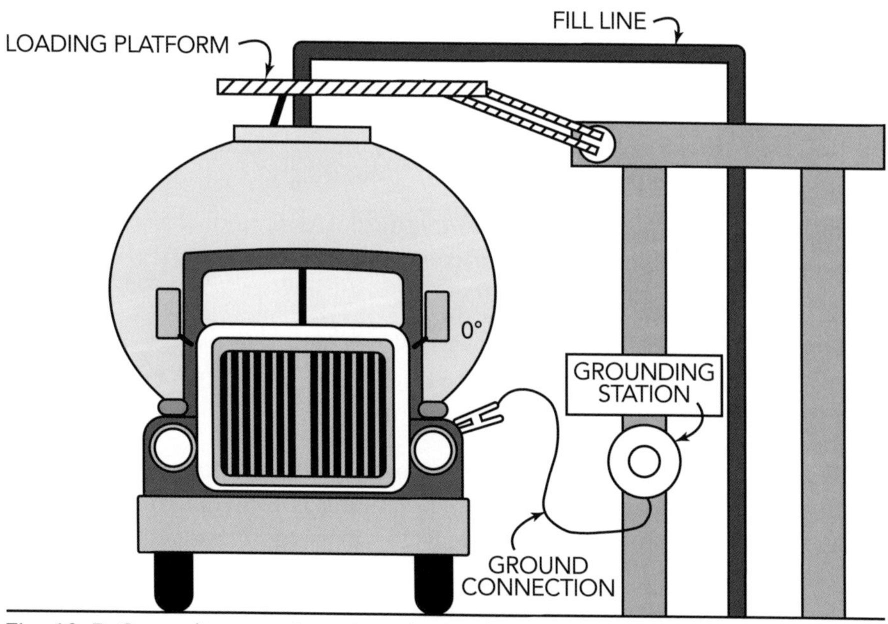

Fig. 10–5. Grounding a tank truck to the loading station

Railcars. The tracks that railcars run on have their own grounding system. The wheel assemblies have metal-to-metal contacts so they are always grounded. But many railcars have wheel bearings that are not conductive, which makes the rest of the carriage assembly isolated electrically. Special connections have to remedy that.

The wear pads on the carriage can also isolate the assembly from the tank and fittings. So the procedure calls for grounding the tank to the loading/

unloading rack during the transfers. Again, the system should have failsafe circuitry that shuts down transfers of fuel if the ground connection is lost.

Delivery sites. Propane tanks at distributors' depots and for residential and commercial locations have the same hazards. Drivers connect their trucks to ground connection on the receiving tanks before starting transfers. The hoses used for transfer are grounded to the truck and therefore to the stationary tank.

Bottles and cylinders. The NFPA recommends that the containers should be bonded (connected) to ground with a clamp with steel points that can penetrate any paint or corrosion products.

People. Operators that handle hoses, valve handles, and so on, can become an important source of static electricity as they move around. In many hazardous sites, operators wear static-dissipating shoes to keep themselves from accumulating static electricity. Static-dissipating flooring also helps.

Maintenance

This subject is joined at the hip to safety and even operations. Bulk plants that handle propane and butane need even more attention to the integrity of the mechanicals than other fuels. Any failure can have more spectacular consequences then, say, gasoline or diesel fuel incidents.

The San Juanico disaster

On November 19, 1984, residents of San Juanico, Mexico, began smelling odorized gas at about 1:00 a.m. Gas had been leaking from a transfer line at the adjacent, huge liquefied petroleum gas (LPG) depot owned by Pemex. Depot staff eventually detected the leak but, unable to find it, fled the premises. At 4:50 a.m., the cloud of gas reached the flare at the plant's edge and ignited. In four minutes, the first LPG sphere erupted in a BLEVE, followed by 12 more explosions in the next hour.

The conflagration incinerated the town of San Juanico and its residents. Over 500 people died, and more than 5,000 suffered severe burns. Large metal tank remnants had landed 4,000 feet away. The investigation ultimately blamed poor maintenance and a faulty gas detection system.

San Juanico remains the largest LPG disaster (fig. 10–6).

Fig. 10–6. Tank remnants from the San Juanico explosions (courtesy of Norwegian University of Science and Technology)

Bulk plants generally have a checklist for operators or inspectors to use to rigorously check the scores of risky situations in a plant. Here are just a few items that need close attention, just to give a flavor of what is on those lists.

Recordkeeping

Documentation of inspections, repairs, and modifications of equipment is essential for several purposes:

- Guides to future change orders
- Timing of future inspections
- Diagnosing subsequent incidents

Corrosion control

As at bulk plants, tanks, piping, and fittings need attention for corrosion prevention. That includes paint or coatings and sometimes insulation. In many places, piping runs underground. Cathodic protection prevents against low-level, ambient electrical currents in the Earth. These can attack piping or, for that matter, anywhere steel meets the ground. Most tanks, spheres, and bullets sit on concrete pads, but some may be partially buried. They need adequate coating or insulation and cathodic protection as well. Even nonmetal surfaces (polymeric handles, gauge housings) may need coatings to protect from weather and exposure to chemicals.

Equipment integrity

Some frequently used or emergency equipment needs special attention. Hoses are prone to kinking and abrasion from dragging. That requires frequent inspection of soft spots, bulges, excessive wear, or cracking that can lead to leaks. Operators have been known to drop hoses, so couplings can be bent or dented and bolts can come loose or missing. Coupling or hose slippage can occur. Frequent testing of the operability of all emergency shutoff valves, pressure relief valves, and alarms is vital. Infrequently used equipment (for emergencies) needs periodic inspection.

Training

Pumps, piping, and fittings for ethane, propane, and butane need different procedures from other liquid hydrocarbons. Everyone involved in handling, modifying, or operating equipment needs continual training, not just at safety meetings. Many companies underwrite remedial sessions, training to upgrade skills, and certification programs.

Government cooperation

Many companies, large and small, have close relations with state, county, and local fire and police departments to prepare for the contingencies. Cooperation with these institutions and with competitors in the area includes establishing communication networks; preparing and executing joint exercises for various types of hazardous material and situations; identifying equipment and human resources available, as well as other resources; and training (fig. 10–7).

Fig. 10–7. Local firefighters training (courtesy of Booneville Fire Department)

Takeaways from Safety, Operations, and Maintenance

- All NGLs are hazardous, but the mobility of propane and butane, the manual handling of transfers, and their road and rail transport make them exceptionally vulnerable to incidents of great consequence.

Exercises

1. What seem to be the root causes of hazardous incidents involving LPG?
2. Who is responsible for safe loading and unloading of NGLs?

CHAPTER 11

Profiles in NGL Enterprise

The only place where success comes before work is in a dictionary.

—Vidal Sassoon (1928–2012)

Contrary to popular lore, Charles Goodyear did not accidentally spill sulfur into a kettle of molten natural rubber and exclaim, "Eureka! I have found it!" He labored for 20 years, searching for a way to make natural rubber durable under hot and cold conditions because he was dismayed that fire hoses cracked in the winter and shoe soles became sticky in the summer. In and out of jail and bankruptcy, he carried on never-ending experiments until he tried adding sulfur. His serendipitous breakthrough came when he inadvertently left some rubber with sulfur added on a hot stove. When he scraped it off, he found an elastomer that withstood both heat and freezing.

For years after that, Goodyear battled dyspepsia, gout, cash flow problems, and patent litigation, and he never made a commercial success of his vulcanization process.

Forty years after Goodyear's death, F. A. Seiberling borrowed $3,500 and started a rubber company. In homage to the inventor, he called it the Goodyear Tire & Rubber Company, now the world's largest rubber business. No Goodyear descendants were ever associated with the enterprise.

Natural gas liquid (NGL) companies that do all the activities painfully covered in the last ten chapters come in many sizes and persuasions, as they have from the beginning: big and small exploration and production (E&P) companies, pipeline companies, liquefied petroleum gas (LPG) distributors, refiners, petrochemical companies, and more recently the so-called midstream companies. Here we trace the history of three—Phillips 66, Enterprise Products, and Texas Star Propane—companies that cover almost all of it.

A Tortuous Past

Phillips 66 came to its position today through a byzantine route, rich with venerable oil industry names and rife with corporate mergers, acquisitions, spinoffs, and joint ventures.

Beginnings

Because of the merger between Conoco and Phillips in 2002, some of the present Phillips 66 has a legacy in the founding of the Continental Oil Company in Ogden, Utah, in 1875, a refining operation committed to supplying kerosene in the western United States.

That same year, E. W. Marland, a man with an almost forgotten, if not forgettable, name, struck oil on land leased from the Ponca Indian tribe in Oklahoma. He built a small refinery there to process the crude, plus hundreds of tiny gas stations to sell the gasoline. In 1929, J. P. Morgan, Jr., gained control of Marland Oil Company. J. P. sacked E. W. and then purchased the Continental Oil Company and set up headquarters in Ponca City. The company adopted the Continental Oil Company name. It kept the distinctive red triangle, a logo that identified the company, until it dropped it in 1970 (the triangle, but not the color red.) Continental even had a petrochemical arm for a while, but later, to concentrate on oil, it sold that business to E. F. Hutton and a group of Conoco executives, who renamed that company Vista Chemicals.

Despite the geographic challenge of its headquarter location, Ponca City, Oklahoma, Continental eventually went global in both the upstream and downstream, including operations in Europe, Libya, and Dubai. All the while, Continental built refineries in the United States and gas stations throughout the world to move the product. Because of its roots, Continental focused its United States operations mainly on markets outside the East Coast and in rural areas.

In one of the several frantic periods of oil industry restructuring, in 1981 DuPont won a bidding contest for Conoco ownership, trumping Seagram's' attempt at takeover. DuPont and Conoco parted ways on friendly terms in 1999 when Conoco went public again in a $4.4 billion initial public offering (IPO), the largest such offering up to that time. (Dotcoms had not been invented yet.) Conoco does not reenter the story for several more years.

The Phillips Brothers

In 1905 Frank Phillips and his brother L. E. drilled a wildcat well on an 80-acre tract in Indian Territory, Oklahoma. That well, the first of 81 successive and successful producers, gushed 250 barrels per day (fig. 11–1). Flush with cash, Frank Phillips purchased a patent from a failing businessman, Walter Snelling. The patent covered a process for distilling propane and butane from the condensate produced at natural gas wells. In 1917, Frank and L. E. consolidated the various holdings they had accumulated into the Phillips Petroleum Company to take advantage of a current oil price increase of 150% (from 40 cents to a dollar a barrel) as World War I started.

Fig. 11–1. The Phillips brothers' first oil well. This producer on the Anna Anderson Lease in 1905 came in at 250 barrels per day.

Natural gas at that time had not much of a market because of the disproportionate difficulties of distribution. Most producers flared all of it. Insightfully seizing an opportunity, Phillips Petroleum built the Hamilton Gasoline Plant in 1917 near Bartlesville, Oklahoma. Using compression and cooling, it extracted from its own natural gas a volatile but marketable gasoline product for the burgeoning automobile market. Shortly after, Phillips engineers developed an even better way, the absorption process, to extract the NGLs and separate the propane and butanes from the heavier materials. Seeing the potential for this expanding supply of fuel, Frank Phillips arranged a joint venture with the Tappan Company to develop affordable gas appliances. By 1924, Phillips had become the largest producer and marketer of NGLs.

The downstream. In 1927 Phillips moved downstream on the oil side by building a 1,500-barrel-per-day refinery in Borger, Texas, and opening its first gas station in Wichita, Kansas. (The gasoline actually came from a nearby gas plant, not the refinery.) By that move, Phillips became an integrated oil and gas company. Like many of its competitors, Phillips was trying to hedge the unfailingly volatile crude oil prices and competition by adding their own guaranteed outlets for crude oil—refining and marketing. That strategy would endure for 84 years.

66?

Frank Phillips himself once explained the origins of the 66 in Phillips 66 during a 1938 message to employees. To get into the gasoline marketing business, they needed a bold brand, he said. The day before the meeting to decide on a logo, a company official was returning to Bartlesville, Oklahoma, in a car testing a new Phillips gasoline. "This car goes like 60 on our new gas," the company official remarked. The driver retorted, "60 nothing," glancing at the speedometer. "We're doing 66!"

After hearing of this exchange around the boardroom table, someone asked where this took place. The reply was Route 66 near Tulsa. That did it. A logo was born.

In 1927, Phillips made a logistical leap forward when it received approval of attractive rail rates for LPG cars and made the first rail shipment of butane. The next year it put on the road the first propane bobtail truck. Phillips had just won a patent dispute over the design of the one drum/two valve system for receiving, storing, and consuming propane. That system enabled delivery of hundreds of gallons of propane to consumers' 250 gallon tanks: no lugging metal bottles back and forth, less frequent deliveries, and more reliable supply. To induce customers, Phillips began leasing residential and commercial storage tanks to save customers the front-end expense. All that opened up the residential heating market. By 1937 it had added to its retail offerings a full line of ranges, refrigerators, and water heaters.

In their ongoing commitment to the business, Phillips executives H. Emerson Thomas and George Oberfell, together with Mark Anton of Suburban Propane Company, founded the predecessor to the National Propane Gas Association. Thomas became the second president.

In 1942, at the behest of the U.S. government, Phillips eased into the petrochemicals business by building plants to turn normal butane into butadiene and natural gas into carbon black. Those two essential ingredients of synthetic rubber were desperately needed for the World War II effort. In 1953 company researchers patented a process for producing polyethylene, and in 1959 Phillips started its first large petrochemical complex on the Houston Ship Channel. It used NGLs as feedstock to an olefins cracker to make the ethylene, the feedstock for their new polyethylene plant. That began another enduring commitment, this time to the NGLs-to-plastics value chain.

To extend its reach, Phillips started in 1952 batching propane in its products pipeline from the Borger, Texas, refining and gas-processing complex to St. Louis and Chicago.

With the upstream opening up NGL supply sources in the Middle East, Phillips partnered with Japan's Bridgestone to venture into international NGL markets. Starting in the 1960s, Phillips chartered, then built long-haul pressurized and semirefrigerated propane carriers. Later this group pioneered the first fully refrigerated very large gas carrier (VLGC), the *Bridgestone Maru*. Phillips became one of the largest participants in the LPG trade among the Middle East, Japan, South Korea, and China. To exploit the LPG opportunities in Mexico and Central and South America, in 1950 Phillips built the first propane export terminal on the Houston ship Channel.

Disappearing musical chairs

The decades following 1980 were ones of daunting corporate and organizational turmoil. In one notorious episode, a freewheeling operator, Tom O'Malley, began buying the unwanted refineries that were being shed by both large and small owners. In 1987 O'Malley and his partners at the financial firm, Argus, bought their first refinery stake, 26% of the Tosco Corporation, a large independent U.S. West Coast refinery. O'Malley stepped in as chairman. In the next 11 years, he perfected his acquisition strategy: buy big distressed refineries, mostly from majors, at a fraction of the replacement cost; increase operating rates and efficiencies; slash costs, a step not at all popular with the entrenched unions. (In some cases, he dismissed all the refineries employees.) As O'Malley said, "Playing hardball is not going to make everybody happy, and sometimes you just have to."[1]

At one point O'Malley's portfolio had 11 discarded refineries from Shell, Unocal, Sun Oil, Exxon, BP, and others. When Phillips bought Tosco from O'Malley in 2001, Phillips added 1.3 million barrels per day to its previous capacity of 355,000 barrels per day. Phillips became the second-largest U.S. refiner behind Valero, and larger than Exxon, Shell, Chevron, BP, Texaco, and Mobil.

Meanwhile, in 2000 Phillips had merged their petrochemicals business with Chevron's in a joint venture called Chevron Phillips Chemical Company (CPCC). Phillips also engaged in a joint venture with Duke Energy Field Services, a natural gas and NGL midstream operation, to form an enterprise that ultimately would handle almost all Phillips's NGL business—the gas plants, the pipelines, trading, and marketing—as DCP Midstream (DCP). During a period of industry horizontal consolidation, in 2002 Conoco and Phillips merged their entire companies. Nine years later, integrated companies were splitting themselves vertically. In 2011, Conoco Phillips split into two entities, with the upstream named ConocoPhillips, and the downstream called Phillips 66. Refining and marketing and the joint venture CPCC went with Phillips 66, as did DCP, the NGL company. By that time, Duke Energy, the JV partner, had been subsumed into a successor firm, Spectra Energy, a natural gas pipeline–based company.

For Phillips 66's NGL business, what has emerged from this upheaval (as of this writing):

- 12 refineries producing, among other things, propane and consuming butanes and natural gasoline.
- Half interest in CPCC, the world's 12th largest petrochemical company, with billions of pounds of NGL-based/ethylene-based petrochemicals.
- Half interest in DCP, a natural gas and natural gas liquids midstream company and the largest U.S. producer of NGLs. Added to that is the propane sales coming from Phillips 66's 12 refineries, making it the largest U.S. marketer of that product.

For Phillips 66 and its ancestral managements, NGLs had always been big, but often overshadowed by much larger businesses—crude oil, natural

1 Wakin, Daniel J. "Private Sector: From Black Monday to Black Gold." *New York Times, February 11, 2001.*

gas, and the refined products gasoline, diesel, jet fuel, and so on. Eventually, they figured out how to get executive focus on NGLs (and petrochemicals)—put them in entities not preoccupied by bigger, more fryable fish. That gave Phillips 66's affiliates, CPCC and DCP, the opportunity to attract their own management talent, capital, and focus.

Elephants Don't Pirouette

To do a pirouette, a ballet dancer must balance on one leg while turning, position the other leg and the hips in the retire position (one leg bent so the toe touches the knee to make the number four), while the arms offset the force of gravity and the head moves to avoid dizziness. Burly people just cannot do it.

Large oil and gas companies have a penchant for looking for the larger oil and gas plays, sometimes called elephants. Often, success demands investment in the infrastructure necessary to exploit their finds: the natural gas processing plants, pipelines to get the production to refining centers, and pipelines and terminals to get the products close to customers. Then they have to deal with the ongoing marketing of the oil, gas, and NGLs.

In the latter part of the 20th and the early part of the 21st century, the biggest oil companies, the majors, and the so-called large independents found numerous opportunities in the United States and around the world to spend money on prospective elephantine projects. As they high-graded their investment opportunities by expected returns, small but still profitable investments in the infrastructure of their existing businesses fell toward the bottom of the list. This was particularly true for NGLs. In many cases, enhancing NGL assets was not considered strategic, or, in the parlance of the day, "mission critical."

Enter Dan Duncan

In the early 1960s, Dan Duncan, a U.S. Post Office worker, studied business and finance at Massey Business College and South Texas College and then joined Wanda Petroleum, a big player in NGLs, as an accountant. In 1968 he left Wanda, and with two partners (whom he shortly thereafter bought out) and $10,000, he started Enterprise Products with two propane transports and a cold call list. In the ensuing 35 years, Duncan built an aggressively entrepreneurial team of businessmen and financiers to springboard his modest holdings to a Fortune 100 company.

A chronological list of events is too intense to detail. It includes both organic growth and acquisitions. Take pipelines as an example. Enterprise bought all or pieces of these: Chinchilla, Stingray, Sailfish, Moray, Ocean

Breeze, Neptune, Nemo, Port Neches, Tri States, Piceance, Rio Grande, Stateline, Fairplay, and Belle Rose pipelines.

While those high five- and six-figure acquisitions went on, Enterprise made billion-dollar transactions: in 2002 the Mid-America Pipeline Company (MAPCO) that connected the Conway-Bushton-Hutchison area of Kansas (the storage and processing hub for NGLs produced in West Texas, Oklahoma, and Kansas) to Minnesota and Chicago; in 2008, the Dixie Pipeline from the Gulf coast to Georgia; and in 2009, the Texas Eastern Products Pipeline Company (TEPPCO) that ran from the Gulf Coast to Chicago and Selkirk, New York. These three pipelines represented the main arteries of the industry's propane distribution east of the Rockies in the United States.

Meanwhile, Enterprise moved into gas processing (fig. 11–2). It built and acquired gas plants and NGL fractionators along the Gulf Coast, in South and West Texas, and in Colorado, connecting them to their hub at Mont Belvieu, where it built and repeatedly expanded its NGL fractionators and leached out, bought, or leased dozens of salt dome caverns for storage.

Fig. 11–2. Enterprise Products' gas plant at Yoakum, Texas. Processing nearly a billion cubic feet per day, the plant produces about 111,000 barrels per day of NGLs.

Seeing the connection between the heavier NGLs and the refining business, Enterprise built methyl tertiary butyl ether (MTBE) plants, butane isomerization plants, and isobutylene fractionation to use the readily available C_4's. It started up C_5/C_6 isomerization plants to enhance the octane number (and value) of natural gasoline. It extended NGL pipeline connections to virtually all the refineries along the Gulf Coast from Texas City, Texas, to Pascagoula, Mississippi.

Petrochemical ventures

Building propylene "concentrators" (see chapter 8) in Louisiana might seem an incongruous stretch for Enterprise, but refinery-grade propylene has about 40% propane with its 60% propylene. Petrochemical companies want 92–95% propylene (chemical grade) or 99.5–99.9% propylene (polymer grade) for their processes. Enterprise knew as much as anyone about fractionating light hydrocarbons, so majors like Exxon Chemical outsourced operations to Enterprise. Having built these "concentrators," Enterprise acquired export terminals on the Gulf Coast to handle both propylene grades. It bought the Lou-Tex Propylene Pipeline, which connected the Houston Ship Channel with Mont Belvieu and south Louisiana. It followed that with propane and butane export capability and then with a fleet of barges for intracoastal transportation.

Natural gas

By 2001, Enterprise had started to back-integrate into natural gas pipelines, including major purchases such as Acadian Gas from Shell Oil (1999) and Gulf Terra from El Paso Natural Gas (2002), as well as some of the pipelines listed above. Even further upstream, it bought natural gas gathering systems that connected wells and fields to gas plants or gas pipelines. Prominent were the Encimal, Canales, and Great Divide Systems.

In 1999 Enterprise made a deal with Shell Oil to process Shell's natural gas production from the Gulf of Mexico. It was not long before Enterprise began thinking of an offshore processing facility (Independence Hub in 2002) and a subsea pipeline (Independence Trail) to bring production ashore. In a joint venture, Enterprise built the Cameron Highway Offshore Pipeline System (CHOPS), which sits along the Outer Continental Shelf of the Gulf of Mexico. CHOPS acts as a gathering system for deepwater crude oil production, and that put Enterprise well into offshore and crude oil pipeline operations.

Structure

Over the years, Enterprise has changed corporate structure like a chameleon traversing a Persian rug, but with purpose. Alliances, mergers, spinoffs, and consolidations all had three objectives—profit, tax efficiency, and control by the Dan Duncan family. The underlying framework was always the master limited partnership (MLP) rather than the corporation. An MLP combines the tax benefits of a limited partnership with the liquidity of common stock. Corporations have to pay income taxes on earnings and then pay out (after-tax) dividends on which the shareholders are taxed again. MLPs have a partnership structure that sells units that are traded just like common stock. They pass untaxed earnings directly to the unit holders who incur the tax liability, but that cleverly (and legally) cuts out one slice of the tax collector's take from the money flow.

Attributes of success

Three attributes of Dan Duncan's enterprise contributed to the remarkable growth of the company: market insight, operational agility, and technical prowess.

- Market insight. Dan Duncan uncannily realized early on that the large, focused oil and gas companies could not or would not pay sufficient attention to opportunities in nonstrategic activities. They would allow and sometimes encourage others to exploit these "small" projects. They would even sell off their own investments in these operations to remove the distraction and to harvest the capital for strategic use.
- Operational agility. The company's business structure allowed swift action, from opportunity identification and development, to deal making, financing, and closing or start-up. Sometimes they would have a half-dozen large acquisitions in the pipeline to be closed in a single year, while still working to grow existing businesses organically.
- Technical prowess. Enterprise built in-house technical competence in a stepwise fashion. Retail propane begat NGL storage, then fractionation, gas plants, gas pipelines and gathering systems, crude oil transportation, offshore processing and pipelines, extension into some NGL-like petrochemicals manufacture and storage, and export terminals and marine transportation.

Pirouettes

The history of the oil and gas business has a recurring theme: major and large independent E&P companies look for, find, and exploit hydrocarbons in places where they are hard to find and produce. To do that, they have to spend money to build the infrastructure necessary to start up and operate. In the years that followed these initial developments, smaller independent operators took advantage of the newly built infrastructure, and whatever technological innovation had been necessary to get the operation going. It has happened in the Middle East, the North Sea, the Gulf of Mexico, and many other parts of the world. And it has happened for a long time in the U.S. NGL business. Enterprise has bought hundreds of millions of dollars of NGL and natural gas assets from Shell, ExxonMobil, Texaco, Dow, Diamond Shamrock, Koch, Sun, Chevron, and others and organized them into a focused business of their own. Other aggressively entrepreneurial companies like Enterprise (Kinder Morgan, Mark West, Ameriprise, DCP Partners, and others) have delicately slipped in between the footprints of those stomping elephants and captured the vacant spaces.

Beyond the Mains

Hundreds of small companies and a few big ones take propane *beyond the mains.* They serve remote service customers who would use natural gas if they could, but for whom the cost of running natural gas mains and distribution pipe all the way to their sparsely populated locations does not make sense to the big natural gas companies.

History

One of these companies, Texas Star Propane (TSP), takes care of customers like these. The owners, Joel Sopchak, Josh Kasprzak, and Gary Hunter, have a long history in propane and related businesses (fig. 11–3). TSP's origins go back to 1950 when Carl Stennett started his retail LPG business in Southeast Texas, selling mostly butane to farmers. By the time his daughter and son-in-law had taken over the business a few decades later, they had switched to all propane. Joel Sopchak, in the third generation, had worked in the Stennett-Sopchak family business for years. Josh Kasprzak had worked the oil fields. Gary Hunter had a long history in trucking and related activities. Together they leveraged Joel's legacy into their own retail propane venture in Magnolia, Texas, starting in 2009. For the first few years, they operated out of Gary's garage.

Operations

TSP now has two commercially presentable locations and operates three bobtail trucks (two 2,700-gallon and one 3,000-gallon trucks) that move over a half-million gallons per year of propane. Magnolia sits about 42 miles northwest of Houston. Despite a scattering of oil and gas–producing operations in the area, they receive their propane supply by transport truck from Mont Belvieu, Texas, an unimaginably complex center for natural gas liquids processing about 75 miles away. TSP's 15,000-gallon storage tank receives loads of 9,000 gallons of odorized propane after it calls them in from one of its three regular suppliers. From that tank, TSP loads its three bobtails to make customer deliveries.

Fig. 11–3. Texas Star Propane's principals

Markets

TSP focused its initial sales efforts in 2009 on residential customers. The Sopchak name had considerable cache and an enduring reputation for customer service. That underpinned TSP's growth quickly to thousands of homes, both in subdivisions and on isolated farms. TSP makes a pledge to every customer to maintain continuity of supply or the next load is free. And that load would be delivered in hours, even on New Year's morning, if need be.

To support growth, TSP maintains a continuous residential solicitation program by mail, incentivized customer-referral programs, and radio advertising on a local station. The three drivers, who are at this point still Joel, Josh, and Gary, regularly stop when they see a potential commercial customer; they assume their sunny sales personas and solicit the business. About 15% of TSP's sales come from commercial accounts—school systems, light industrial, restaurants, body shops, retail stores, forklift operators, and cylinder refill stations.

To maintain their service reputation, TSP drivers keep meticulously detailed records on deliveries. Too small yet to warrant elaborate computerized systems yet, they track the drawdown rate by season of each customer in a separate paper file, so they can time the frequency of refills. Residential customers typically have tanks of 250 gallons, which may require a drop-off every 90 days during April through October, but every 20 days in the winter. Large homes, 8,000–9,000 square feet, might have bigger (500-gallon) tanks, but they still need deliveries twice as often. Commercial customers typically have tanks that hold 500–1,000 gallons.

TSP provides propane tanks to its customers for lease or purchase if the installation is aboveground. TSP's own technical associate will handle the placement and connections. Some customers, mostly residential, would gaze askance at a tank in their backyard and so prefer an underground installation. Buried tanks have different metallurgy and coating and require special corrosion protection, plus the heavy equipment to place them in the ground. TSP leaves the extra installation effort (and considerable expense) for the customer to bear.

Safety and regulations

The Texas Railroad Commission (TRC) reaches its tentacles well into retail propane distribution. Truck and terminal operators have to attend operating safety schools, be tested, and renew their certification every four years. On new installations, tanks, lines, meters, and safety devices have to be inspected by TRC agents, in order to prevent mechanical failure or operator error that could lead to fire or explosion. TSP hires outside contractors to conduct regular safety classes with mandatory attendance by all eight employees.

The company gives every new customer an initial integrity inspection of the tank, the lines, and every appliance that draws propane at the customer site, which could include furnaces, heaters, stoves, clothes dryers, water heaters, and outbuilding space heaters. When a tank inadvertently goes empty, besides refilling it for free, TSP will reinspect the whole system again make sure it was not a leak that caused the inopportune drawdown. That is a TRC regulation, but with this procedure, TSP tries to turn this lemon of a burden into a marketing tool: reassuring the customer of TSP's attention to the system's integrity and the customer's welfare.

As is the case with many other exurban propane dealers, demographics favor a bright future for TSP. For years, agricultural area populations had declined as people migrated to the cities, looking for whatever the restless look for—jobs, culture, excitement. Now cities are sprawling, and light industry desperately searches for low-cost land and labor outside of urban areas; residential and commercial development moves out just ahead of industrial development. All that promises propane dealers a growing market *beyond the mains.*

Takeaways from Profiles in NGL Enterprise[2]

- Phillips 66's approach is one big oil company's way to bring focus on and commitment to opportunities in the NGL business.
- The history of Enterprise Products reflects how an entrepreneur can build a company from scratch.
- Texas Star Propane, the quintessential small dealer, looks to a bright future.

Exercises

1. What is the common theme across Phillips 66, Enterprise Products, and Texas Star Propane?

2 Note: Information about Texas Star Propane comes from interviews with its principals. Much of the history of Enterprise Products comes from its website and research into secondary sources. The same goes for Phillips 66, plus insights from *Phillips: The First 66 Years*, by Robert Finney. Bartlesville, OK: Phillips Petroleum Company, 1983.

Glossary of NGL-Related Terms

absorber. Refinery or gas plant unit that absorbs light gases (or perhaps acid gases) from a natural gas stream or other streams and makes the gases available for further processing as a separated stream.

alkylation. Refinery unit that reacts isobutane with propylene and/or butylenes to make a high-octane gasoline blending component called alkylate.

API. American Petroleum Institute.

aromatics. Chemical molecules that have a benzene ring in their make-up.

associated gas. Gas co-produced with crude oil from an oil well.

ASTM. American Society for Testing Materials.

atomic number. The number of protons in an atom.

atomic weight. The weight of an atom relative to the weight of carbon-12, one of the carbon isotopes.

autoLPG. Same as autogas.

autogas. Propane and/or butane used for automobile fuel.

batch system. A pipeline set up where one commodity is loaded and shipped in a pipeline followed by another (batch).

benzene. C_6H_6, a cyclic compound with six carbons in a ring, found in petroleum, made in refineries and olefins plants, and one of the base petrochemicals.

bitumen. 1. Asphalt (British). 2. Very heavy crude oil produced by heating underground resources to make it viscous enough to pull out of the ground.

BLEVE (boiling liquid expanding vapor explosion). An explosion that may occur when propane or butanes leak uncontrollably.

bobtail. A handy size truck that can transport LPG to retail customers.

branch. In a hydrocarbon paraffin chain, a methyl group attached to an internal carbon.

Btu (British thermal unit). The measured amount of heat needed to raise 1 pound of water 1°F.

buffer. In a pipeline, a smaller amount of a liquid in between two batches that keeps them from downgrading each other.

butadiene. C_4H_6, a chemical created in a cracking process; used to make rubber.

cat cracker. A refinery unit that cracks heavy gas oil into light products.

cat reformer. Refinery unit that upgrades the octane number of naphtha to a high-octane gasoline blending component called reformate.

catalyst. A substance that can enhance a chemical reaction by speeding it up, lowering its pressure or temperature, or reducing the number of side reactions.

cavern. Underground storage leached from a salt dome or mined in rock.

Celsius. Measure of temperature named after Swedish astronomer Anders Celsius (1701–1744).

coker. A refinery unit that cracks very heavy oil (vacuum flasher bottoms) into light products and coke.

condensate. Oil coming from a separator at a natural gas field that has the characteristics of a very light crude oil.

cracking. Severing the bonds that hold hydrocarbon molecules together, such as ethane cracking into ethylene and hydrogen. Also, the operative reactions in refinery cat crackers, hydrocrackers, and cokers to make smaller molecules out of larger ones.

cricondentherm. Temperature on a natural gas phase diagram beyond which natural gas is assuredly all vapor.

critical point. Temperature on a phase diagram that divides the all-liquid state from the all-vapor state.

cyclic compound. Hydrocarbon in which the attached carbon (or other) atoms form a closed loop.

cylinder. A portable, 2- to 10-gallon metal container for propane or butane.

dehydration. Removal of water vapor from a natural gas stream.

dehydrogenation. Removal of a hydrogen atom from a hydrocarbon molecule.

desulfurization. Removal of sulfur atoms from oil or gas.

dew point. The temperature at which heavier vapors contained in lighter vapor condense and drop out as liquid.

distilling unit. Refinery unit that separates crude oil into several fractions.

dry gas. Natural gas that (1) has had the water vapor removed or (2) has had most of the NGLs removed.

erosion. Wearing away of the Earth's surface.

Fahrenheit. Measure of temperature devised by physicist Daniel Gabriel Fahrenheit (1686–1736).

fracking. Hydraulic fracturing, a technique for creating pathways for oil and gas to flow by creating cracks perpendicular to a well in subsurface formations such as shale.

fractionator. Distilling column that separates liquid chemicals from each other using heating, vaporization, cooling, and condensing.

gas. Sometimes (in the United States) it means gasoline; sometimes it means natural gas.

gas chromatography. Measurement technique for determining the composition of a gas that uses a process where individual compounds give off an emission at different, measurable, and identifiable times.

GPA. Gas Processors Association.

HD-10. Commercial propane specification (heavy duty—10% propylene content) in California.

HD-5. Commercial propane specification (heavy duty—5% propylene content).

heat exchanger. Device that allows the transfer of heat from one stream, liquid or vapor, to another.

heavy liquid. In an olefins plant, either naphtha, natural gasoline, or gas oil feedstock.

HHV (higher heating value). A measure of the heat given off when a hydrocarbon is burned.

higher heating value. See HHV.

hydrocarbon. Molecule of carbon and hydrogen atoms.

hydrocracker. A refinery unit that adds hydrogen as it cracks gas oils into light products.

hydrodesulfurization. Refinery process unit in which sulfur atoms are removed by reacting compounds with hydrogen over a catalyst.

hydrotreater. Refinery process that reacts hydrogen with oil streams to remove sulfur atoms and other contaminants.

isomer. Molecule with the same number and types of atoms as another molecule but that has a different structural configuration.

isomerization. Refinery unit that upgrades normal pentane and hexane to isopentane and isohexane or converts normal butane to isobutane.

lean oil. Solvent used in an absorption process.

LHV. Lower heating value, which takes into account the energy necessary to turn the water created by combustion from liquid to vapor.

LNG (liquefied natural gas). Predominantly methane at a temperature to keep it liquid.

LPG (liquefied petroleum gas). Propane and/or butane.

mercaptan. Family of sulfur compounds that has the group SH bonded to a carbon.

Merox. Proprietary process for removing sulfur, mostly used for gasoline-type streams.

migration. Movement of oil and gas from the source rock where it was created toward the Earth's surface.

mole fraction. Measure of the percent of molecules or atoms of an element in a mixture.

molecular weight. Relative weight of an atom or molecule measured by the atomic weight of the atoms involved.

MON (motor octane number). A measure of octane under conditions of stress or heavy load.

naphtha. Oil that boils at temperatures slightly higher than gasoline and slightly lower than kerosene.

naphthene. Refiners' general name for C_5 to C_8 cyclic compounds.

NGL. Natural gas liquid.

nonassociated gas. Gas from a gas well (not an oil well).

NPGA. National Propane Gas Association.

octane. Measure of the compression ratio at which a gasoline sample will spontaneously preignite.

odorant fade. The declining effectiveness of some odorants over time.

organic compounds. Compounds consisting of carbon and hydrogen and perhaps other atoms.

packed column. Distillation column that has packing, that is, shaped objects or corrugated sheets or other material on which condensation and vaporization can take place.

PERC. Propane Education and Research Council.

phase diagram. Graph of the temperatures and pressures at which various physical states of a chemical can exist: solid, liquid, and vapor.

pig. Pipeline injection gadget used to buffer different batches or to clean the pipeline or to check the integrity of the pipe.

pygas. Gasoline stream from an olefins unit cracking naphtha, natural gasoline, or gas oil.

pyrolysis. Use of high temperature to induce a chemical change, as in an olefins plant.

raw make. Same as Y-grade.

reboiler. Attachment to a distilling unit that reheats the bottom stream and sends the resulting vapors back into the distilling column for redistillation.

reflux. Attachment to a distilling unit that cools the top stream and sends the resulting liquid back into the distilling column for redistillation.

reservoir. A sedimentary formation with sufficiently porous rock that can hold an accumulation of oil, gas, and/or water. Needs an trap above to contain any oil or gas.

rich gas. Natural gas before the NGLs are removed.

RON (research octane number). A measure of octane under conditions of mild driving.

RVP (Reid vapor pressure). A measure of vapor pressure used in the petroleum business.

salt cavern. A cavity that has been leached out of a salt dome by pumping freshwater in, dissolving salt, and pulling brine out.

salt dome. A large accumulation of salt that has oozed up from a subsurface body of salt left from the vaporization of a saltwater sea.

saturates. Generally butanes, propane, and ethane.

saturates gas plant. Refinery unit that separates normal and isobutanes, propane, ethane, and methane from each other.

seismic. Technique for mapping subsurface characteristics by bouncing sound waves off sedimentary layers and recording the returned signals.

stabilizer. Distilling column that removes butanes and lighter gases from a stream.

sweetening. Removal of sulfur compounds from a natural gas stream.

thermal cracker. A refinery unit that cracks heavy oil (vacuum flasher bottoms) into light products. Most thermal crackers have been replaced by cokers.

toluene. $C_6H_5CH_3$, a cyclic compound with six carbons in a ring plus a methyl group attached to one, found in petroleum, made in refineries and olefins plants, and one of the base petrochemicals.

trayed column. Distilling column that has multiple, perforated trays on which the condensation and vaporization can take place.

turboexpander. Unit in a gas plant that reduces the temperature of the gas low enough such that most of the NGLs liquefy and can be separated in a distilling column.

unsaturates. Generally butylenes, propylene, and ethylene.

valence. Measure of the number of atoms to which an atom has the propensity to bond itself.

vapor lock. Gasoline vaporizing in a car's gasoline line from the gas tank to the engine, thus starving and stopping the engine.

vapor pressure. The amount of pressure needed to prevent vapors from emanating from a liquid.

VLGC (very large gas carrier). An ocean tanker that transports liquefied LPG.

volatiles. Sometimes interchangeable with NGLs, but specifically meaning liquid hydrocarbons easily vaporized at room temperature and pressure.

wet gas. Natural gas that (1) has noticeable water vapor or (2) has not had the NGLs removed.

Wobbe index. Measure of the burning characteristics of natural gas.

xylenes. $C_6H_5(CH_3)_2$, cyclic compounds with six carbons in a ring plus two methyl groups attached to two of the carbons (there are three isomers, three ways to attach them), found in petroleum, made in refineries and olefins plants, and one of the base petrochemicals.

Y-grade. In a pipeline, a mixture of NGLs.

Answers to the Exercises

Chapter 2

1. After several serious explosions (e.g., Ardmore, Oklahoma; New London, Texas; Hindenburg zeppelin), governments and industry looked at the root causes and set standards and regulations to prevent future accidents.
2. The internal combustion engine; Snelling's bottling propane and butane; development of the LPG railcar; Smokey Billue's salt cavern; Curme's ethane cracking process for ethylene.
3. Ocean-going tankers.
4. In World War II all the isobutane was used in an alkylation plant to make aviation gasoline components; the conversion of butane to butadiene to make rubber was developed; gasoline was rationed, and so farmers turned to propane for tractor fuel.

Chapter 3

1. Source rock (with the transformed microorganisms), migration, reservoir, and trap.
2. The permeability of shale (the ability of fluids to move through the rock) is so low that the rock has to be fractured to make channels for the fluids to move at commercial rates.
3. Drill pipe, which has been pulled out, casing, production casing, production tubing, and miscellaneous devices for perforating or fracturing.
4. In most cases, the oil, gas, and water coming from the well go through a field separator; oil goes to a tank; gas goes into a gas-gathering system; and water goes into a tank for further disposal or reinjection into the subsurface.

Chapter 4

1. Propane, propylene, normal butane, isobutane, normal butylene, and isobutylene. (There are actually two normal butylenes, *cis*-butylene and *trans*butylene, slightly different because of the orientation of the terminal methyl group.)
2. Natural gasoline from each source is made up of a unique combination of molecules that determine its unique physical properties.
3. The sulfur compounds to be taken out of propane are too high in concentration to meet environmental and other regulatory specifications. The sulfur put back in is an odorant compound, which facilitates leak detection and is far lower in concentration than the sulfur compounds taken out.

Chapter 5

1. Natural gas pipelines have requirements for sulfur content, carbon dioxide content, and water contents, all of which demand that almost all produced gas be treated to protect the pipeline and the customers. Economically, if the price spread between the NGLs and natural gas is sufficient, which it often is, it makes sense for the gas plant to separate the two for sales in their respective markets.
2. Most new gas plants of any size use turboexpander technology rather than absorption or refrigeration.
3. Sweetening refers to the removal of sulfur compounds. Conditioning, and sometimes treating, refers to the removal of all the contaminants, acid gases, water, and nitrogen.

Chapter 6

1. No chemistry takes place in separation units such as distilling. Whatever molecules go in, come out, albeit separated in groups from one another.
2. They do not have to be different except that the proportion of methane that a natural gas plant handles as a percent of the total is much more, 80%–95%. In a refinery gas plant, methane accounts for about 10%–20% of the feed. And, of course, in many refineries, the olefin-containing gas liquid streams are separated at the alkylation plant where the olefins are *reacted* out and the propane and butanes are then separated by *fractionation.* So the size of the compressors, distillation columns, and turboexpanders in the three types of plants are quite different.

Chapter 7

1. Refineries receive butanes by truck, railcar, pipeline, barge, tanker, and in diluted crude oil. And, of course, they make their own in various units.
2. A refinery can build a butane isomerization (BI) plant to convert normal butane to isobutane.
3. Ethane can be shipped as a liquid if its temperature can be kept below 90°F, the critical temperature. Above that, it is a critical gas; propane and the butanes are shipped under pressure as liquids. Natural gasoline moves as a liquid.
4. In order of increasing costs per gallon of construction: salt dome caverns, rock caverns, steel storage.

Chapter 8

1. Ethane: 1×10^9 lb ethylene ÷ 0.80 lb of ethylene/lb feed = 1,250,000,000 lb

Propane	0.40	2,500,000,000 lb
Butane	0.36	3,600,000,000 lb
Natural gasoline	0.23	4,350,000,000 lb

2. Ethane: 1,250,000,000 lb ÷ 2.97 lb/gal ÷ 42 gal/bbl ÷ 365 = 27,454 bbl/day

   Propane: 2,500,000,000 lb ÷ 2.50 lb/gal ÷ 42 gal/bbl ÷ 365 = 38,462 bbl/day

   Butane: 3,600,000,000 lb ÷ 4.24 lb/gal ÷ 42 gal/bbl ÷ 365 = 55,385 bbl/day

   Nat. gasoline: 4,300,000,000 lb ÷ 5.40 lb/gal ÷ 42 gal/bbl ÷ 365 = 51,944 bbl/day

3. In the short run, the Totally Flexible Chemical Company should crack less propane and more butane or natural gasoline, if the economics warrant it. If, despite its name, its olefin plant cannot handle heavier feed, the alternative in the long run is to build a PDH unit.

Chapter 9

1. $10{,}000 \times 6.0 + 52X = 9 \times (10{,}000 + X)$
   $X = 697.7$ bbl butane
2. | | |
   |---|---|
   | 1,000 bbl cat cracked × 4.4 = | 4,400 |
   | 2,000 bbl reformate × 2.8 = | 5,600 |
   | 1,000 bbl alkylate × 4.0 = | 4,000 |
   | 400 bbl ethanol × 18.0 = | 7,200 |
   | Total: 4,400 | 21,200 |

   $21{,}200 \div 4{,}400 = 4.82$ RVP

   $4{,}400 \times 4.82 + 52X = 9 \times (4{,}400 + X)$

   $X = 427.7$ bbl of butane
3. Advantages: lower emissions, good cold starts, cleaner burning, high octane.
   Disadvantages: fewer fueling stations and service personnel scarce in some countries, trunk space limitations, more weighty fuel tank, more hazardous.

Chapter 10

1. Human error, equipment failure, and acts of God (lightning, floods, etc.).
2. The driver/loader has the first line responsibility to perform the work safely. Management of the organization that employs the driver/loader has the ultimate responsibility and achieves this through training and safety programs for all employees.

Chapter 11

1. They all have been organized to focus on NGLs without the distraction of any other parts of the oil and gas business.

Dew Point and Wobbe Index

Many folks observe dew points on autumn mornings. As the overnight temperatures drop, the air can no longer keep all the water vapor (humidity) dissolved in vapor form. Water goes from vapor to tiny droplets landing on windshields and giving a glistening effect to lawns. Natural gas producers typically remove most of the water vapor from their transmitted gas to eliminate the chance of liquid water formation in a pipeline. This dehydration step is often done using a hygroscopic material (one that has an affinity for water) such as glycol. That reduces the risks of corrosion and the formation of potentially damaging liquid slugs in the pipelines both the gathering systems and the main natural gas pipelines.

Natural gas pipeline companies set specifications for the hydrocarbon dew points of the gas delivered to them by producers and gathering system companies. They want to ensure that the hydrocarbon gases heavier than methane are not in such high concentrations that they might condense—turn from vapor to liquid—as they travel down the pipeline. For this reason, they watch the gas dew points.

Since most pipeline companies also receive natural gas from multiple sources, they want to make sure that the resulting mixtures will meet the specifications of their customers' burner tips, both residential (furnaces, hot water heaters, clothes dryers, and stoves) and commercial (burner tip and industrial boilers and other equipment). For this they watch the *Wobbe index* (explained later in this appendix).

Dew Point

The plot of hydrocarbon dew point for a pure gas (e.g., methane or propane) is a single line on a pressure-versus-temperature chart. The higher the pressure, the higher the dew point, the temperature at which gas turns to liquid. But natural gas, as it leaves the producing lease, is almost always a combination of methane plus some natural gas liquids (NGLs) and water, all in the vapor state. For complicated reasons, this kind of mixture behaves unlike a pure gas: At some temperatures and pressures, the fluid can be a liquid, a gas, or a combination of gas *and* liquid in a stable state. A phase diagram, as in figure A1–1, shows the relationship.

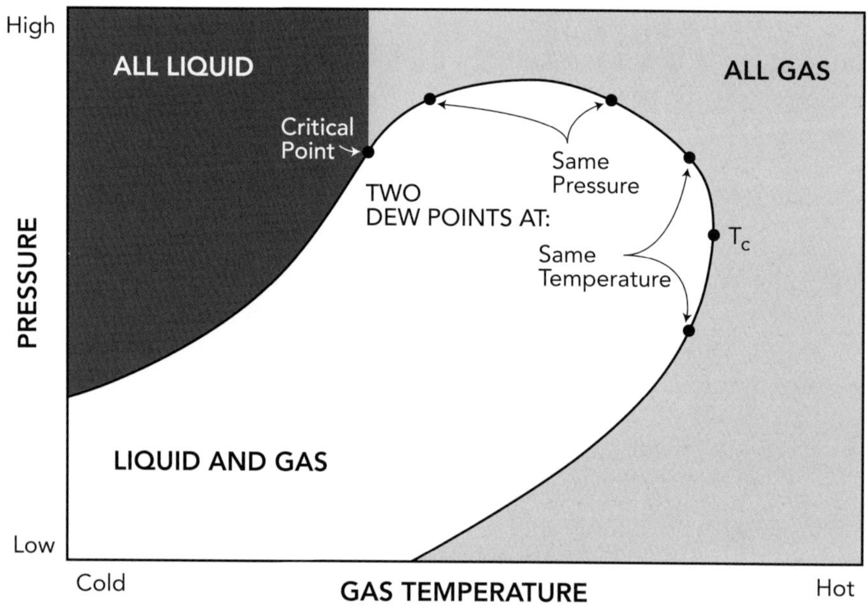

Fig. A1–1. Dew point diagram of natural gas with some NGLs in it

The natural gas stream operating at a high enough temperature, to the right of T_c, will always be a vapor. At combinations of pressure and temperature in the darkened area, the gas will be all liquid. (Natural gas pipelines do not operate there.) In the solid white area, various proportions of gas and liquids coexist. The size and shape of the solid white area depends on the content of the NGLs. Pipelines often operate in this region. The more NGL and the less methane, the bigger the area under the curve.

Figure A1–1 has a few additional features:

- At some pressures, a dew point can occur at two temperatures.
- At some temperatures, a dew point can occur at two pressures.
- There is a temperature, T_c, called the *cricondentherm*, beyond which natural gas is all vapor.

Cricondentherm comes from *critical condensation temperature* and entered into general use in the 1950s as gas pipelines went national in the United States. As in chapter 4, the more NGLs in the natural gas, the higher the heating value and the energy content. In that vague way the dew point has a rough correspondence to heating value, but the relationship is not a quantitatively operational tool.

Figure A1–2 shows the effect of processing wet natural gas at a gas plant, removing some of the NGLs, with the gas/liquid area, temperatures, and pressures shrinking toward the origin.

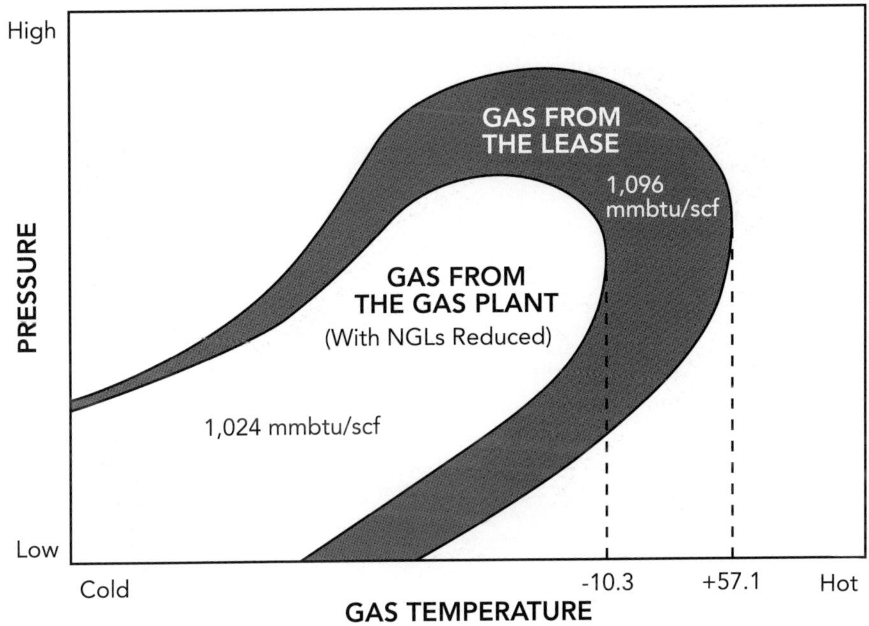

Fig. A1–2. Impact of reducing the NGL content of wet gas

Wobbe Index

Before natural gas became commercial, starting in the 19th century, pipeline companies supplied town gas. They manufactured this gas mostly from coal. It had an energy content of 600–800 Btu per standard cubic foot (Btu/SCF), which is about 60–80% of today's natural gas. Consumers used town gas for heating, lighting, and cooking, and sometimes for heating.

As the supplies of natural gas became available and pipelines reached new markets, town gas could not compete and dried up. In the switchover from town gas to natural gas, every burner tip had to be replaced because the higher energy content of natural gas, more than 1,000 Btu/SCF, needed a different amount of air mixed with it to burn efficiently. The effort was huge.

The pipelines and local distribution companies, motivated by pressure from the equipment manufacturers, started to pay more attention to how natural gas behaved at the burner tip or in turbines. They noted that higher NGL content could lead to the following:

- Lift-off, where the flame becomes detached from the burner tip and returns with a sputter or even flameout, a dangerous phenomenon
- Flashback, where the flame burns *inside* the burner tip instead of outside, which could eventually destroy the burner tip (fig. A1–3)
- Incomplete combustion, as evidenced by a yellow-tipped flame, which is a waste of energy

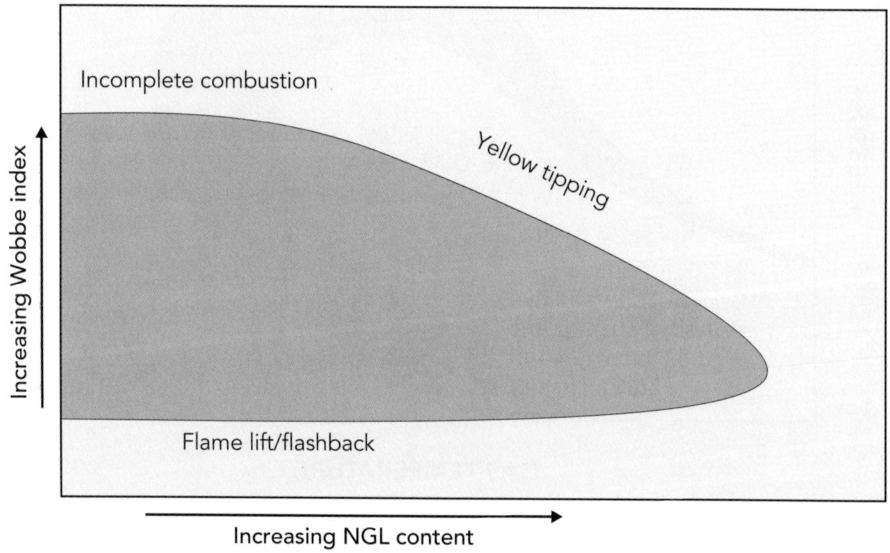

Fig. A1–3. Impact of NGL content on natural gas burning characteristics

In 1927, Goffredo Wobbe quantified the relationship between efficient burning and gas properties, naming it the Wobbe index. Many refer to it also as the Wobbe number. He had found that efficient and effective burning was related to the energy content of the gas divided by the square root of its specific gravity:

$$WI = HHV \div \sqrt{sp.\ gr.}$$

where WI is the Wobbe index

HHV is the higher heating value (see chap. 4)

$\sqrt{sp.\ gr.}$ is the square root of specific gravity of the gas (relative to air)

The way the mathematics work out, HHV and WI are both measured in Btu/SCF because specific gravity is a ratio, a dimensionless number. That could be confusing except that the Wobbe index is usually quoted as just a number and HHV usually has the dimensions after it. In the United States, for example, the Wobbe index of most natural gas falls in the 1,250–1,400 range, while the HHV may run 980–1,180 Btu/SCF. Equipment manufacturers design their burner tips to accommodate these properties.

As the NGL content of natural gas increases, the Wobbe index increases, as evident in table A1–1. That gives rise to considerable discussion in commercial and regulatory circles about setting standards for the Wobbe index. Many believe it is the correct measure for ensuring the interchangeability of various natural gas streams—from gas plants, LNG sources, and, increasingly, NGL-laden shale gas.

Table A1–1. Higher heating value (HHV) and Wobbe index

	HHV (Btu/SCF)	Specific Gravity	Wobbe Index
Methane	1,012	0.55	1,365
Ethane	1,713	1.04	1,739
Propane	2,522	1.52	2,046
Butanes	3,255	2.00	2,301
Pentane (a natural gasoline surrogate)	4,020	2.49	2,547

Natural gas streams are now routinely monitored for heat content and specific gravity using gas-liquid chromatograph technology. That allows calculation of the dew point and Wobbe index in real time.

Treating Gas at Gas Plants

Besides methane, ethane, propane, butanes, natural gasoline, and condensate, many other things come out of a gas well, usually in small percentages, though in the case of water, it could be as much as 50%. Most of these contaminants have to be removed to make the natural gas good enough for injection into the natural gas grid. Figure A2–1 shows a general processing scheme in a gas plant for typical amounts of everything.

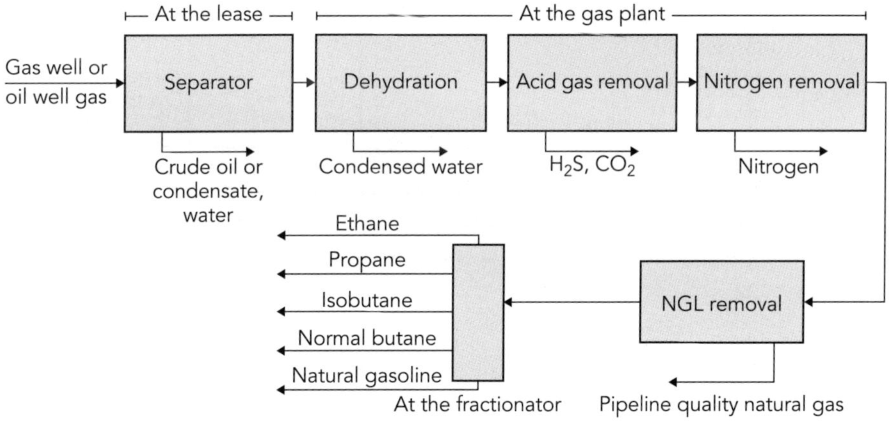

Fig. A2–1. General processing scheme in a gas plant

Removing Water Vapor

After the heavier oils (condensate or crude oil) and water are separated from the natural gas or natural gas liquid (NGL) stream at the well site, there still remains the likelihood of entrained water vapor in the gas. The water content of natural gas going into a pipeline has to be less than the pipeline's specification (often less than 7 pounds water per MMcf). Otherwise, the water could condense and accumulate in low spots or even form icy hydrates (a slushlike mixture of water and methane, and maybe ethane, propane, and butane) and constrain flow. Water combined with a small amount of entrained acid gases can cause a slow but corrosive reaction with almost anything it touches.

A glycol dehydration system is most often used to remove the water from the gas stream. It can be located at the well site or gas plant. The gas flows into the bottom of an extraction unit, a column with perforated trays, baffles, or packing (fig. A2–2). Liquid triethylene glycol (TEG) or a similar chemical enters near the top of the column. TEG is hygroscopic; that is, it has an affinity for water. It extracts the water vapor from the gas and carries it out the bottom of the column. Dehydrated gas, sometimes called dry gas, exits the top of the column.

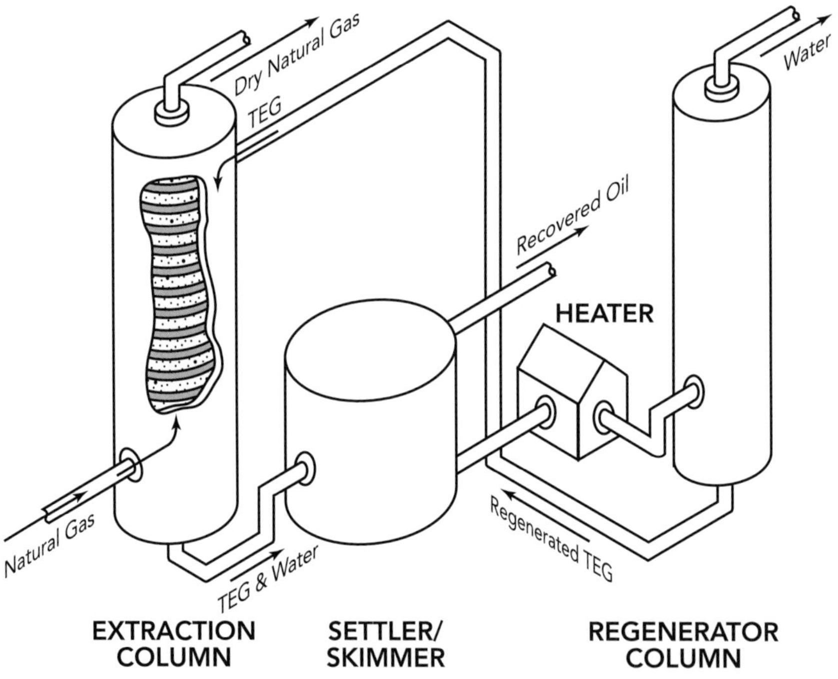

Fig. A2–2. Dehydration

The TEG from the bottom then goes to a regeneration step. First, it goes to a separator where any entrained oil can be skimmed off. Then it goes to a heater where the temperature is raised to about 400°F. This causes the water (only) to evaporate, leaving behind the TEG, whose boiling point is 549°F.

The regenerated TEG goes back to the top of the extraction column for reuse. The water vapor goes either to the atmosphere or to a cooler to condense any hydrocarbons that may have made it through the entire process.

In an alternative dehydration process, a desiccant or adsorbent like a zeolite molecular sieve or silica or alumina gel in two or more parallel steel vessels captures the water. The gas stream is cycled among the multiple vessels. While one vessel removes the water, in another vessel that has been saturated with water, a warmed slipstream of the already-dried gas is passed through to carry off the entrained water in a much reduced concentration. When that is completed, the wet natural gas stream is switched back into the dried-out vessel.

In a few places, refrigeration or membrane systems do the dehydration.

Removing Contaminants

Bad actors, particularly the *acid gases* (hydrogen sulfide [H_2S] and carbon dioxide [CO_2]) are endemic in natural gas operations. H_2S combines with water to form sulfuric acid, which will eat into metal, destroying machinery, instruments, and pipelines. Likewise, CO_2 forms carbonic acid, which does the same thing. Other sulfur compounds such as carbon disulfide (CS_2), carbonyl sulfide (COS), and mercaptans might also be present.

The most common process for removing the acid gases from the natural gas stream at a well site or main gas-processing plants uses amine treating. The chemical family of amines includes tongue-twisting names, which is the reason they are referred to by their acronyms: monoethanolamine (MEA), diethanolamine (DEA), methyl diethyl amine (MDEA), diglycolamine (DGA), diisopropanolamine (DIPA), and triethanol amine (TDA). These compounds have the nearly magical ability to absorb CO_2 and the sulfur compounds without taking in natural gas or NGLs. Most plants use DEA these days, particularly for H_2S removal from the whole gas stream before fractionation. However, there are varying benefits and downsides to the different amines in initial cost, effectiveness in removing particular sulfur compounds, energy use in regeneration, and degradation in service.

Some plants use a potassium chloride carbonate solution rather than an amine to perform the same function. At others, diisopropyl alcohol is used for COS removal from the fractionated propane stream. A proprietary process called Merox, a catalyzed caustic stream, is commonly used to remove mercaptans from the propane and butane products.

The Treater

The amine-treating system looks much like the dehydration system in figure A2–2, without the settler/skimmer. Natural gas is introduced to the bottom of a column with baffles, packing, or perforated trays. As the gas rises by gravity, it meets the liquid amine, which has entered near the top of the column. As the gas and the amine slosh past each other, the amine absorbs almost all the CO_2 and the sulfur compounds.

The treated gas exits from the top of the column with a much reduced content of the acid gas. The acid gas–laden amine exits the bottom of the column and goes to a second column operating under a slight vacuum. With a little heating on the way in, the acid gases vaporize out of the amine solution and exit from the column top. From the bottom of the column, the regenerated amine exit goes back to the top of the absorber column to be reused.

The acid gases go to another unit (usually close by), typically a Claus plant, which converts the H_2S to elemental sulfur. The other sulfur compounds, COS, CS_2, and mercaptans, if in sufficient presence, might go to a SCOT (Shell Claus Offgas Treatment) unit for conversion to H_2S so they can be sent also to the Claus unit.

Sulfur from the Claus unit comes out molten. Some places store it hot. Others lay it out in forms and let it cool and solidify into slabs or flakes. Sulfur can be transported in any of these forms, depending on the needs of the customer.

Some natural gas streams are outliers with very little acid gas content. In some of those cases, the natural gas might pass through a membrane system. The acid gases can make it through the membrane, but the natural gas and NGL molecules are too big and cannot. The acid gases are captured leaving the treated, sweet natural gas stream. A few natural gas streams can have huge H_2S content—as much as 40%—requiring large extraction columns and sulfur plants.

APPENDIX 3

Gas Plant Yield and Shrinkage Calculations

Ethane comes into a gas plant in a gaseous state, just as the other natural gas liquids (NGLs) and methane do. It leaves the gas plant in a liquid state, just as the other NGLs (but not methane) do. Unfortunately, for anyone keeping track of the intakes and outturns, the ratios of the volumes of the ethane, propane, butanes, and natural gasoline in the gaseous state are different than their ratios in the liquid state. That is why engineers have to use an unfamiliar measure, *mole percent*, to handle the balances.

Moles

In the 18th and 19th centuries, scientist Amedeo Avogadro and his contemporaries pieced together how gas and liquid measurements work. Avogadro discovered that, astoundingly, equal volumes of gas, at the same temperature and pressure, contain the same number of molecules. He found that the number of atoms or molecules in a mole of gas is always 6.22×10^{23}. The finding was so remarkably profound that his contemporaries gave his name to it: Avogadro's number.

A century later, two German physicists, August Krönig and Rudolph Clausius, derived the ideal gas law, $PV = nRT$, and Dmitri Mendeleev found out what the constant, n, in the equation is. All this gave the fundamentals for calculating how much NGL *in gallons* should come out of a gas plant when the mole percent composition of the gas going in is known.

But what is a mole of gas?

> Definition: A mole of gas (called a gram mole when dealing in grams and liters) is equal to the atomic weight of the gas in grams. The atomic weight of helium is 2, so that a gram mole of helium is 2 grams (and has 6.2 × 1023 molecules in it).

So where do atomic weights come from? Over the centuries scientists have painstakingly measured the atomic weights of all known elements as they discovered the elements. The periodic table of elements shows the abbreviation for each element's name and its atomic weight. Here are some examples of atomic weights, in round numbers: hydrogen is 1, helium is 2, carbon is 12, and oxygen is 16.

That makes a gram mole of water, H_2O, weigh 18 grams, the sum of the parts (2 + 2 + 16 = 18) and of methane, CH_4, weigh 16 grams.

Gas to Liquid Calculations

Gas plant engineers use a handy device, a gas chromatograph, to measure the gaseous composition of wet gas coming into a gas plant. The results come out in gram mole percent. Engineers use mole percent instead of volume percent to measure the amount of certain components in natural gas. The volume of gas varies with temperature and pressure, so it is difficult to quantify the percent volume of NGLs in natural gas. Mole percent defines how many NGL molecules are in 1 mole of natural gas, which will not change with temperature and pressure.

In the United States, engineers work in pounds and gallons, not grams and liters. So they need to do a couple of transformations between those two systems to get the answer in the terms they want—the volume in gallons (not liters) coming out of a gas plant if 100% of the NGLs are (theoretically) recovered.

To explain their procedure, use the typical wet gas composition in table 4–4 in chapter 4 and 1 MSCF (thousand standard cubic feet) of wet gas.

Conversion from Mol % (Mole Percent) to GPM (Gallons per MSCF)

To understand the process of unit conversion, follow along in table A3–1.

Line 1 is the molecular weight (to a couple of places) of each NGL. Line 2 is a mole in grams, and Line 3 is a mole in pounds, having divided the mole in grams by the number of grams in a pound. For ethane it is:

$$\text{pound mole of ethane} = 30.07 \text{ grams} \div 465.59 \text{ grams/pound}$$

Table A3–1. Calculation of the wet gas content of NGLs in gallons

	ethane	propane	isobutane	normal butane	natural gasoline*
1 molecular weight	30.07	44.097	58.124	58.124	72.151
2 grams per mole	30.07	44.097	58.124	58.124	72.151
3 pounds per mole	0.0662	0.0952	0.01255	0.1255	0.1558
4 pounds per gallon	2.96	4.22	4.86	4.86	5.22
5 gal/mole [(3) ÷ (4)]	0.02232	0.023	0.0263	0.0273	0.302
6 gal/MSCF [(5) ÷ 0.000836603]	26.68	27.49	32.64	31.46	36.13
For a 150 MMSCF/D plant					
7 mole % in wet gas	3.40%	1.50%	0.40%	0.50%	0.60%
8 gal per MSCF [(6) x (7)]	0.907	0.412	0.131	0.157	0.217
9 Mgal/D in 150 MMSCF/D	136.1	61.8	19.7	23.6	32.6

* An estimate of the C_5^+ fraction, which contains C_6's, C_7's, and more.

Line 4 is the weight of a liquid NGL gallon at standard pressure and temperature. Line 5 is the number of gallons in a mole of each of the NGLs, which is as follows for ethane:

gallons per mole of ethane = 0.0662 lb mole ÷ 2.96 lb/gallon
= 0.02232 gallons/lb mole

Line 6 goes back to fundamental physical chemistry. It might need a chemist to explain it well, but that is beyond the scope of even this appendix. It might even seem like the situation in figure A3–1. At any rate, based on that gas equation and its constant devised by Krönig, Clausius, and Mendeleev in the 19th century, the volume of 1 mole of gas at standard condition is as follows:

$$V \text{ per mole} = nRT/P = 1 \text{ mole} \times [8.314 \text{ J/(mol} \times \text{K)}] \times 288.7 \text{ K}/101{,}325 \text{ Pa}$$
$$= 0.02368996 \text{ m}^3$$
$$= 0.000836603 \text{ MSCF}$$

where 8.314 is the universal gas constant in Joules per mole degree, 288.7 K is temperature in Kelvins and 101,325 Pa is the standard unit of pressure measured in Pascals, both being at standard conditions

So, if 1 pound mole of gas takes up that little space, 0.000836603 MSCF, then Line 6 calculates the number of gallons per MSCF of each NGL. For example, for ethane in Line 6:

gallons of ethane per MSCF = 0.02232 ÷ 0.000836603 = 26.68 gal/MSCF

Finally, given a handy size gas plant processing, say, 150 MMSCF/D, of wet gas, if 100% of the NGLs were recovered, which they seldom are, than Line 7 gives the purported output of the NGLs. For ethane:

content of ethane = 0.097 gal/MSCF × 150 MMSCF/day= 136.1 Mgal/day

as shown in table A3–1, along with the other NGLs.

ethane 150,000 MSCF/day × 0.907 = 136.1 Mgal/day
propane 150,000 MSCF/day × 0.412 = 61.8 Mgal/day
isobutane 150,000 MSCF/day × 0.131 = 19.7 Mgal/day
normal butane 150,000 MSCF/day × 0.157 = 23.6 Mgal/day
natural gasoline 150,000 MSCF/day × 0.208 = 31.2 Mgal/day

"DOBBS. I THINK YOU NEED TO BE MORE EXPLICIT HERE."

Fig. A3–1. NGL calculation, inspired by the cartoons of Sidney Harris

Shrinkage

Removing NGLs from wet natural gas results, of course, in a reduction of the natural gas volume. That reduces the revenue associated with the natural gas stream just as it generates the new NGL revenue stream. The calculation of the difference between these two, called the shrinkage value, is just as complicated as everything else in this appendix.

To get those values, note that natural gas is priced and sold by its heat content, $/MMBtu. That calls for calculating the shrinkage using the heat content of both the methane and the NGLs (using the higher heating values [HHVs] in table 4–3 in chapter 4.)

Start with Row 1 of table A3–2, the higher heating values. Next in Row 2 is the wet gas feed composition (same as in Line 7 in Table A3–1) in mole %.

Row 3 has the targeted percent recovery of the NGLs from the wet gas, 90% of the ethane and so on.

Table A3–2. Calculation of the wet gas heat content of NGLs

	Methane	Ethane	Propane	Isobutane	Normal butane	Natural gasoline	Total
1 (HHV) Btu/SCF	1,010	1,780	2,516	3,354	3,369	4,700	
2 Feed mole%	93.0	3.4	1.5	0.4	0.5	0.6	
3 NGL recovery %		90	98	99	99	100	
4 Residue gas calculation	93.0	0.35	0.03	0.004	0.005	0.0	93.379
5 Residue gas mole %	99.6	0.39	0.03	0.004	0.005	0.0	100.0
6 Feed gas Btu/SCF	939.3	60.5	37.7	0.0	16.8	28.2	1,096.0
7 Residue gas Btu/SCF	1,006.0	6.5	0.8	.01	.02	0	1,013.5

(Rounding involved)

To get the gas composition of the residue gas (feed gas after the NGLs are removed), row 4 has the original 93.0% mole of methane in the feed gas and the unrecovered NGLs. For ethane that would be 3.4% from row 2 times 10% (i.e., 100–90% recovery), equaling 0.35. The sum of row 4, that is, 93.379, is the weighted average of the residue gas, a number that will figure next.

Row 5 has the numbers reflecting the calculation of the residue gas on a 100% basis, not the 93.379% in row 4. Each row 5 entry equals the row 4 entry divided by the weighted average, 0.93989, the decimal version of 93.979. That allows row 5 to add up to 100. Ethane for example, equals 0.035 ÷ 0.93979 = 0.36; the sum is 100%, allowing for rounding.

Rows 6 and 7 get the heat content of the feed gas and the residue gas in Btu/SCF by multiplying rows 2 and 4 (changed to decimals) by row 1 and summing. For the ethane entry,

$$\text{in feed gas MMBtu/SCF} = 3.4 \times 0.01 \times 1{,}780 = 60.5$$
$$\text{in residue gas MMBtu/SCF} = 3.6 \times 0.01 \times 1{,}780 = 6.5$$

and the weighted average of the feed gas is 1,082.3 Btu/SCF and the residue gas is 1013.7 Btu/SCF.

Finally, the shrinkage value for the gas plant removing the NGLs:

volume of feed gas times the Btu content of the feed gas
minus the volume of the residue gas times the Btu content of the residue gas
times the price of natural gas in $/MMBtu, say, $3/MMBtu

The volume of the residue gas has shrunk to 93.979% of the 150 MMSCF/D feed (from the sum of row 4) or 150 × 0.93379 = 140.07 MMSCF/D. So,

$$\text{shrinkage} = [(150 \text{ MMSCF/D} \times 1{,}096) - (140.07 \times 1013.7)] \times \$4$$
$$= \$89{,}767/\text{D}$$

That ought to be less than the sales revenue for the NGLs removed from the gas if the gas plant is to be making money by taking the NGLs out of the feed gas.

Octane Number Measurement

Posted on the gasoline pumps at every gas station is an octane rating for each grade of gasoline. In the owner's manual for every car, in the section on fuel, the manufacturer recommends the octane number best suited for that car. What does all that mean to the car owner? What is an octane number anyway?

Understanding octane numbers requires a working knowledge of how internal combustion engines work. Figure A4–1 gives the overall workings of a four-stroke engine. (Some people casually call it a four-cycle engine, but as will become apparent, there are only two cycles—the piston goes up and down just twice.)

The Engine

These are the essential parts of the internal combustion engine:

- Cylinder
- Piston
- Intake and exhaust valves at the top
- Spark plug
- Connecting rod between the piston and the shaft

Enter the cycle where the intake valve is allowing a mixture of air and gasoline vapor into the cylinder while the piston heads toward the bottom. The valve closes, and the piston starts up the cylinder, compressing the gasoline/air mixture.

As the piston reaches the top of the stroke, the spark plug lights off, emitting a spark that ignites the fuel. The fuel burns almost explosively, expanding as it does, driving the piston down in the power stroke.

At the bottom of the power stroke, the exhaust valve opens, and as the piston is pushed up the cylinder, the combustion gases, mostly carbon dioxide (CO_2) and water vapor, exit. Then the first cycle starts again.

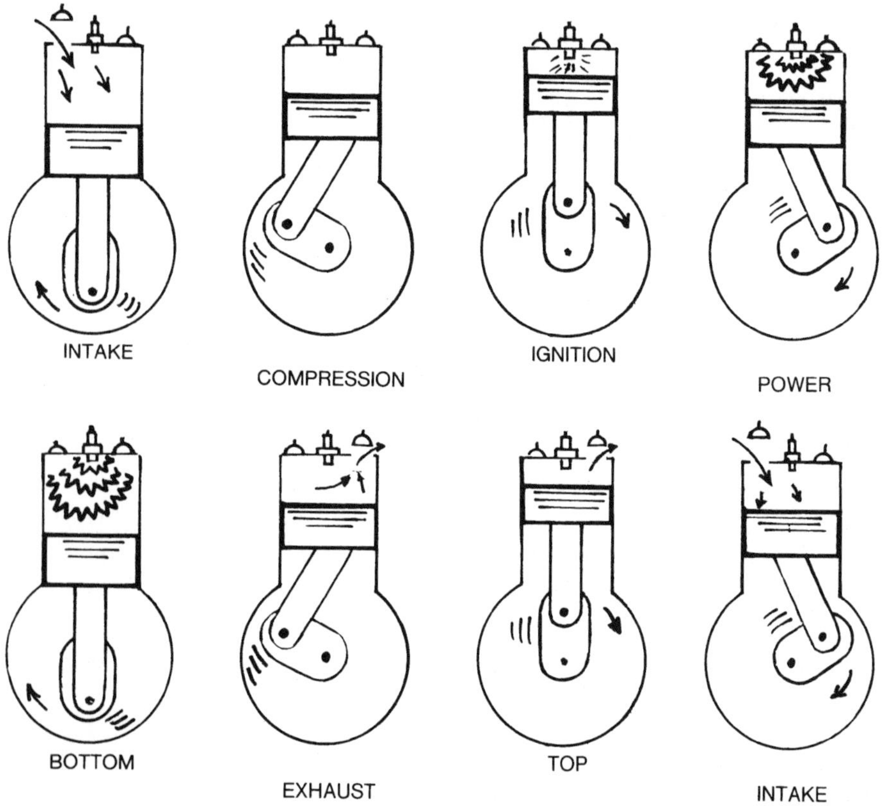

Fig. A4–1. Two-cycle internal combustion engine operation

Preignition

In the 18th century, the French scientist and balloonist, Jacques Charles, conjured up the physical relationship between pressure and temperature so well known today: Compressing the volume of a gas will increase its temperature. Cars run on that principle today. As the piston compresses the fuel/air mixture in the cylinder, the temperature increases proportionately. That's okay, unless the temperature gets so high that the fuel/air mixture ignites spontaneously, before the piston reaches the top of the stroke and before the spark plug lights off. Bad things happen if that occurs.

The driver may or may not hear the preignition as a pinging or chattering in the engine, both called knocking. In the engine, the fuel/air mixture has started to burn and expand, pushing down on the piston as it tries to move up. The connecting rod and shaft receives unwanted stress. Moreover, since the engine is working against itself, the fuel efficiency, that is, miles per gallon, suffers.

The Right Stuff

Different model cars have different engine configurations. Some cars compress the fuel/air mixture a lot—the big cars. Some, not so much. The car manufacturers recommend the kind of gasoline suitable for the compression ratio of every car model, the ratio of the volume when the cylinder is empty to the volume when the piston is at the top of the compression stroke, call it V1/V2. The measure they use is the octane number, which they calculate using that suitability ratio.

Testing for Octane Number

Octane numbers are measured against two somewhat arbitrary standards using a test engine in the "knock" lab. The compound normal heptane, C_7H_{16}, which is very susceptible to knocking, is defined as a zero octane gasoline. At the other extreme, isooctane, C_8H_{18}, is defined as 100 octane. By mixing the two together in exact proportions, a gasoline of any octane between 0 and 100 can be created.

At the knock lab, the tester uses the chart in figure A4–2 to determine the octane number of any gasoline or gasoline component, using the following steps:

1. Run the unknown gasoline sample in a test engine that has a movable head that can change the compression ratio, V1/V2. The tester gradually increases the compression ratio, listening until the engine starting to ping or knock. The tester notes that point on the compression ratio axis of the chart (Point A).
2. Next the tester makes a mixture of gasoline that is below the likely octane number of the component whose octane number is to be determined. (Usually this is easy enough to figure out by testers who are doing this for a career.) In this case, a gasoline with an octane number of 86 is blended by mixing 86% isooctane and 14% normal heptane. By definition, that's 86 octane.
3. Run this mixture in the test engine, again increasing the compression ratio, V1/V2, by moving the head. The tester listens until the engine

starts to ping again, but now the tester has a point that is on *both* the compression ratio and the octane number axes. The tester notes that point on the chart (Point B).

4. The tester mixes a blend of 94 octane material—94% isooctane and 6% normal heptane—and runs that in the test engine using the same procedure. When the engine starts to knock, the tester puts another point on the chart (Point C).
5. Now the tester has the engine calibrated by connecting Point B and Point C.
6. By drawing a line from Point A, the compression ratio of the sample to be determined, to the new line between B and C, and then down to the bottom scale, the tester finds the octane number of the sample, 92 octane.

The blending operation in a refinery takes various components and mixes them to get finished gasoline. To get the right octane number, the octane number of all the components have to be added together in proportion to their amount in the blend.

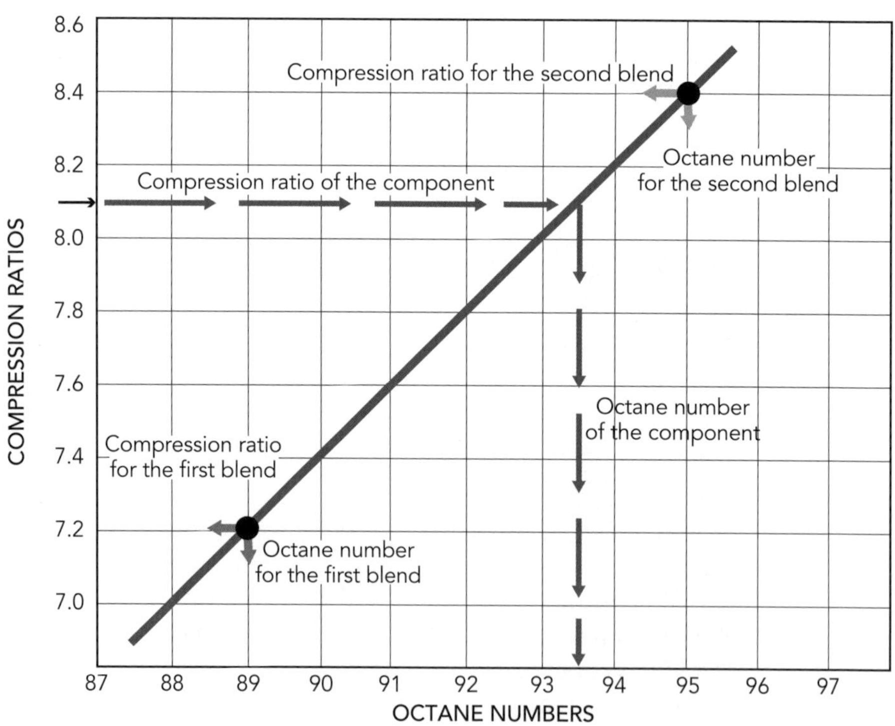

Fig. A4–2. Octane calculation chart

Different Measures

Octane numbers are measured using two different test engine conditions, mostly speed. The idea is to test the gasoline under different driving conditions.

- The Research Octane Number (RON) measures driving under pleasantly normal conditions.
- The Motor Octane Number (MON) measures driving under stressful conditions, such as up a steep hill or pulling a boat.
- In Europe and other parts of the world, the octane number posted on the pump has been the RON. In the United States, the octane rating posted on the pump is the average of the two, (R + M)/2, the octane number shown in figure A4–3.

(R + M)/2 !? What does that mean?

In the 1970s, the U.S. Federal Trade Commission proposed that gas stations start to post the MON in addition to the RON that they had been using. "Silly," said the gas station industry. "Consumers don't even know what RON is, no less MON!"

The automobile manufacturers favored posting just the MON. It gave the impression that their cars could run on lower octane fuel. The Federal Trade Commission (FTC), the gas station people, and the car manufacturers argued for months on end. Finally the FTC issued the regulation: "Post the average, (R + M)/2," which gives a number that didn't mean anything to anyone . . . but the debate was over.

Fig. A4–3. Octane rating on a pump

APPENDIX 5

Ethane and Supercritical Properties

For engineers and operators dealing with ethane and ethylene pipelines, the notion of supercritical fluids is essential to planning, design, and operations. Supercritical fluids defy the normal laws of gas behavior as codified by Jacques Charles in about 1785. He never published his work, but Joseph Louis Gay-Lussac did in 1801. Gay-Lussac was unaccountably ethical enough to deem the relationship Charles's law. Before that (1662), Robert Boyle had published his law, based on the earlier observations of Richard Towneley and Henry Power, but Boyle, unlike Dalton, decided to attach his own name to it. Boyle's law, in simple terms, states:

> *The pressure of a constant mass of gas is inversely proportional to its temperature:*

$$P = 1/V$$

Charles's law states, again in simple terms:

> *The volume of a constant mass of gas is inversely proportional to its temperature:*

$$V = 1/T.$$

It was not long after that, in 1811, that Amedeo Avogadro put it all together with the familiar (at least to chemists and chemical engineers) ideal gas law, which states:

The volume of a constant mass of gas times its pressure is proportional to its temperature:

$$PV = nRT$$

where n is the amount of gas
R is the ideal gas constant

All those laws work most of the time, but not around a phenomenon known as the critical temperature. To explain that, look at the phase diagram for ethane in figure A5–1. (An ethylene diagram looks much the same.)

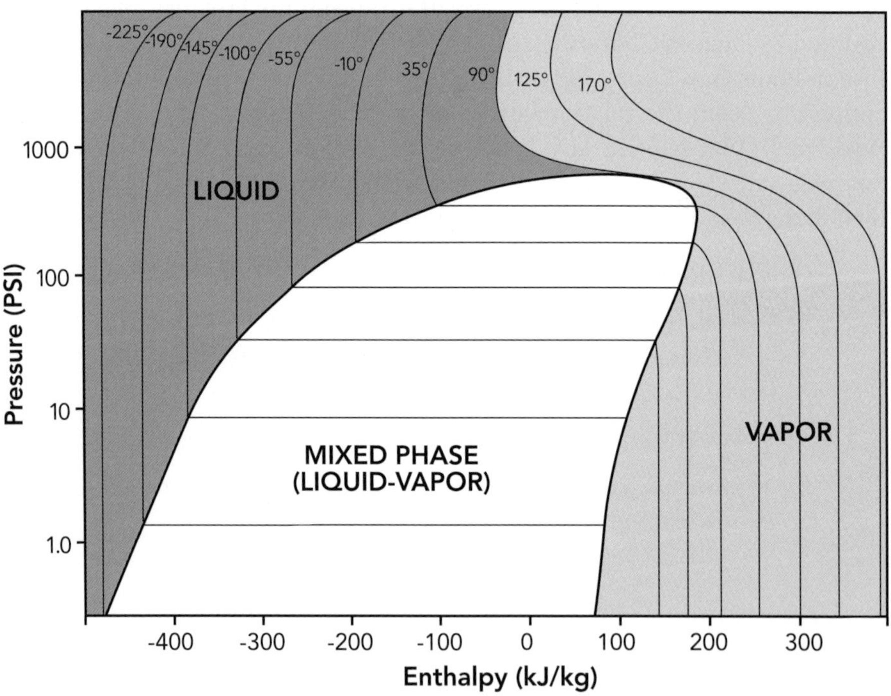

Fig. A5–1. Ethane phase diagram. Critical temperatures show vapor-liquid phase borders.

The axes, pressure, and, strangely enough, enthalpy (which is measured in heat content per pound) make sense when looking at the way temperatures wiggle around the diagram. For this exercise, understanding more about enthalpy is not very important.

Six important points about pipelining ethane (or ethylene) come from this diagram:

- A suite of constant temperature lines, more or less vertical, sweep across the chart.
- Way to the left of the chart, there is a region at low temperatures where *only* liquid exists.
- Way to the right, there is a region where only vapor exists.
- The dividing line is the critical temperature for ethane, 90°F or 32.2°C; for ethylene, 49°F or 9.5°C.
- There is an envelope in the middle where both liquid and vapor coexist.
- Some constant temperature lines pass through the region where there are both liquid and vapor.
- At the top of that region is a constant temperature line, the critical temperature: To the right of that line, only vapor exists; it is compressible but it *does not follow Charles's or Boyle's or Avogadro's law*, and no matter what the pressure, this gas will not turn to liquid in this region!

For an ethane pipeline, ambient conditions usually do not allow keeping the fluid above the critical temperature of 90°F. But at, say, 65°F and 1,000 psi, the fluid is to the left of critical temperature line and is liquid. If the pressure drops at delivery, the conditions may stray into the vapor-liquid region.

Index

C

D

E

F

G

M

N

O

P

Q

R

S

T

U

V

W

X

Y